MRINAL K. DAS

A GUIDE TO FIELD IDENTIFICATION

REPTILES

OF NORTH AMERICA

by
Hobart M. Smith
and Edmund D. Brodie, Jr.

Illustrated by
David M. Dennis
and Sy Barlowe

MRINAL K. DAS

GOLDEN PRESS • NEW YORK

WESTERN PUBLISHING COMPANY, INC.
RACINE, WISCONSIN

PREFACE

The fascination reptiles hold for many people and the fear they inspire in others are both based on the inherent mystery of these secretive, poorly known animals. It is our hope that by facilitating the identification of species, misunderstanding and fear will decrease—and fascination and study will increase.

Reptiles and amphibians comprise an important part of the natural world. To their comprehension and appreciation, and through these to their conservation, this book and its predecessor on amphibians are dedicated. In so dedicating these books we declare our adherance to the concept that self-preservation alone is not the proper goal of mankind; rather, it is the preservation of the whole of our functioning ecosystem, of which amphibians and reptiles are a vital part.

We express gratitude to the countless herpetologists whose contributions have brought the understanding of reptiles to its present level. We thank many colleagues for help in preparing this guide, among them: Kraig Adler, Ronn Altig, Ray Ashton, Walter Auffenberg, Joe Beatty, Joe Beraducci, Robert Brown, James Collins, Joseph Collins, George Dalrymple, James Dixon, Kenneth Dodd, Nick Dye, Carl Ernst, George Fichter, Neil Ford, Richard Funk, Whitfield Gibbons, Michael Goode, David Haynes, Gary Helseth, Donald Hoffmeister, Eric Juterbock, Merritt Keasey, Arnold Kluge, Robin Lawson, Arthur Leitheuser, Michael McCoid, Paul Moler, Richard Montanucci, Henry Mushinsky, Ronald Nussbaum, Lewis Ober, Robert and Arthur Rosenberg, Henry Russell, Raymond Semlitsch, David Stephen, Robert Storm, Wayne Van Devender, and Steve Wylie. Special thanks are due Findlay Russell for his help on the snakebite section.

We deeply appreciate the help of our editor, Caroline Greenberg, whose patience, attention to detail, and talent for making the difficult understandable greatly contributed to the quality of this book. The artistic talent and accuracy of Sy Barlowe (pp. 23-47, 51-63, 67, 71-75, 79, 81, 89-93, 97, 107, 121) and David Dennis (all other illustrations) made our work much easier. We thank our wives, Rozella B. Smith and Judith Johnson Brodie, for their support and tolerance. Brodie is compelled to acknowledge with gratitude Kenneth M. Walker for delivering him from ophidiophobia.

H.M.S. and E.D.B.

TABLE OF CONTENTS

REPTILES—class Reptilia

INTRODUCING REPTILES

This book is a guide to the identification of turtles, lizards, snakes, amphisbaenids, and crocodilians of North America north of Mexico. A total of 22 families, 109 genera, 278 species, and over 500 subspecies are treated. Among them are 18 exotic species that were introduced either intentionally or by accident and have now established resident populations in North America.

Like the amphibians from which they evolved, reptiles are air breathers and have little internal control of their body temperature (ectotherms), but the reptiles have advanced features that make their habitat less restrictive. All reptiles either give birth to their young (viviparous) or lay eggs (oviparous) on land. All reptiles also have an outer covering of scales or scutes enabling them to live away from the moist surroundings to which amphibians are confined. Temperature is thus the limiting factor in the distribution of reptiles as a group.

The diets and food habits of reptiles, as a group, are diverse. Some snakes, for example, feed exclusively on cold-blooded vertebrates; others primarily on worms, grubs, or insects; and still others make their meals mainly of mice or other rodents. Nearly all reptiles are predatory carnivores, or flesh eaters.

As a group, lizards are essentially insect eaters (insectivorous). Most turtles eat fish, frogs, crustaceans, and other small animals that live in or near water, but a few kinds of turtles and lizards eat both animals and plants (omnivorous) or are almost wholly vegetarian (herbivorous), particularly as adults. Those kinds of snakes that make their meals of rodents are obviously beneficial. The benefits of lizard predation on insects are less obvious but are also helpful. All reptiles play a role in the intricate web of life and are thus important elements in their particular habitats and ecosystems.

Nearly all reptiles are wary and secretive, their abundance often much greater than suspected. Most snakes and some lizards and turtles are active primarily at night (nocturnal) or only during evening and morning hours (crepuscular). Most lizards and turtles are active during daylight hours (diurnal). Turtles may bask in the sun but are quick to get out of sight at the slightest commotion. Snakes and lizards generally keep under cover, too, but lizards are the more conspicuous and can often be observed chasing and catching their insect prey. Male lizards may perform openly to attract females, and they chase after other males in territorial squabbles.

Where Reptiles Live

In North America, reptiles are most abundant in the warmer southern regions. Alligators do range as far north as South Carolina and Louisiana in resident populations, but they are by far most numerous in the near-subtropics of southern Florida. Snakes are most abundant both as individuals and in species in the southeastern United States, but some species range northward into Canada. A few species of lizards also occur as far north as Canada, but as a group, lizards are best represented in the warm and dry southwestern areas. Turtles, too, are widely distributed but are most abundant where it is warm. A few species—notably the tortoises—are adapted to dry (xeric) conditions, but as a group, turtles are most abundantly represented either in or near water.

When and Where to Look

In temperate climates, reptiles are dormant in winter. In extreme southern United States, it is warm enough for them to remain active throughout the year. Cool weather slows all reptiles, however, and they do not become active again until the weather warms. Male alligators, and also other crocodilians, roar or bellow in the spring to signal dominance of a territory, and both male and female alligators grunt, uttering a piglike noise that is apparently used as a "call" signal. Turtles, snakes, and lizards make sounds, but there are no vocal congresses to establish mating sites as there are in amphibians. Reptiles thus can be found, with the exception of crocodilians, only by seeing them, and for many species, the finds are as much by accident as by skill.

Knowing where the species lives has considerable bearing on the success of a reptile hunt, of course. To see aquatic turtles either feeding or basking, for example, approach a pond, lake, or stream cautiously. Expose as little of yourself as possible, do not make quick movements, and give all likely spots where the turtles might be sunning your special attention. Sometimes the turtles will bask at the surface with only their heads exposed. Softshell turtles bury themselves in the sand in shallow water and then stick their long, snakelike necks up to the surface, their tubelike snouts serving as snorkels. But even if no turtle is in sight at the moment, perhaps a vibration or some small noise has alerted them. If a turtle was active at a place, it will soon appear again if you wait quietly.

Lizards are the most conspicuous of the reptiles. While they will scurry out of sight—under a rock or log or onto the other side of a tree or limb—you will often hear their rustling even if you missed seeing them at first. Again, if your objective is watching them in nature, stand very quietly or find a good place to sit comfortably. Soon the lizards will be on the move again, accepting you as a part of their surroundings.

Most snakes are active at night and are only seen sunning during the day. It is difficult to observe them going about their daily lives. Aquatic snakes are the easiest of all to find because of their limited habitat. They are always either in the water or nearby. Terrestrial snakes are most easily found in their hiding places—under rocks, logs, bark, debris, or other movable objects. Do not, however, take this as a license to destroy habitats. Whatever is lifted or moved should be put back in place exactly as it was originally.

Collecting

Reptile collecting is done almost totally in the daytime. Turtles are found in their hiding places—under rocks, in debris, or buried in sand or mud. Those that are buried leave small funnel-shaped depressions where they draw in their heads. With experience, you can recognize the approximate size of the turtle by the size of the funnel, then run your hands down into the sand or mud so that one is on each side of the turtle's carapace. Softshell turtles are commonly caught in this manner. Remember that they have extremely long and snakelike necks, sharp jaws, and vicious tempers. The bite of a large softshell can be severely painful, and the turtle is also reluctant to let loose. Snappers can also be dangerous, especially large ones. Be extremely careful in handling them. Turtles do not have teeth, but the edges of their jaws are sharp and their jaw muscles powerful.

All lizards can bite and are quick to do so, holding on tenaciously. Both the Gila Monster and the Mexican Beaded Lizard are poisonous, the only venomous lizards in the world. Their venom is primarily neurotoxic and can cause death. These lizards should be avoided, though they have only a primitive mechanism for injecting venoms and deaths from their bites are extremely rare. Bites of other lizards are harmless, but if the skin is broken, apply an antiseptic to curb any secondary infection. Do not grab a lizard by its tail, for in many species, the tail readily breaks off (autotomizes).

While most snakes are harmless, their bites no more dangerous than those from lizards, there are in North America a number of venomous species: the Copperhead, the Cottonmouth, coral snakes (2 species), and rattlesnakes (15 species). Snakes should not be handled unless you are absolutely sure of their identification. If you are picking up a snake of doubtful identity, use a noose on a long stick or a stick with a short fork to hold the snake behind its head. And pick up the snake by holding it behind its head so that it cannot bite. Hold the snake firmly. Snakes are difficult to restrain and can manage almost uncanny twists and turns. Snakes known to be harmless can be picked up by any part of their body, of course. But even harmless snakes can give painful bites. They strike and often draw back quickly, their back-curved teeth tearing the skin. Some snakes "chew" and are reluctant to release their hold.

Unbleached muslin bags of various sizes, double-sewed with rounded corners to eliminate loose, entangling threads, make excellent containers for reptiles. The bags should be at least twice as deep as wide and should have a tie string two inches (50 mm) or more from the top. These bags are strong and will not break even if moist litter is put in them. They can be carried looped over your belt, which makes them easy to get when they are needed.

Remember that venomous snakes have long fangs and can bite through the bags. Put the bags in a box or other container through which the snakes cannot bite or carry the bags at the end of a long stick or pole. Snappers and other turtles can also bite through a bag. This can be avoided, too, by carrying the bags properly.

Never put collected specimens in the sun. Many snakes succumb because collectors leave the bags exposed. Keep bags in the cool shade.

Ethics

Leaving the habitat as nearly as possible exactly as you found it is a cardinal rule. Do not collect more specimens than you need for your study. Return the specimens alive to where they were collected after you have completed your observations. Because of habitat destruction—highways, housing developments, clearing for farming, and the like—a number of species of reptiles are now threatened with extinction.

Before you go collecting anywhere, check with state and federal wildlife representatives to learn what species are endangered and what can be collected. Never collect without first getting permission from the landowner. A refusal is rare, but if you do not have permission, you may be fined for trespassing, which can be both costly and embarrassing.

Observing

It has been the custom for herpetological hobbyists, and even some professionals, to express their interests by "going collecting"—merely trying to find and bag as many species or specimens (or both) as possible. Even though the specimens may be returned to their home territories later, the lives of many animals are unduly disrupted, often fatally, and seldom is as much learned about them as could be the case by simply observing the animals without disturbing them.

With the present increasing concern for our vanishing wildlife, the pursuit of herpetology should increasingly emphasize field observation with minimal disturbance. Astonishingly little is known about reptilian behavior under natural conditions—certainly a field of investigation that is destined to become popular in the future as students learn to "live and let live."

SNAKEBITE

Bites of nonvenomous snakes can be treated with an antiseptic. The danger from these bites is secondary infection; some bites may warrant giving an antitetanus agent.

Venomous snakes, however, are indeed dangerous. Avoid picking up or handling poisonous snakes. (Of course, if you are collecting, this is sometimes unavoidable.) In areas where poisonous snakes live, wear boots, look before putting either your hands or your feet anywhere, and be very careful when you must cross fences or logs. Follow common sense precautions, and the chances of being bitten are extremely slim. It is also possible to be bitten by a venomous snake and not have poison injected. If you know the snake is venomous, keep calm and plan on how to get to a medical facility as quickly as possible.

The venom apparatus of snakes consists of large modified salivary glands connected by ducts to fangs on either side of the upper jaw. The snake controls the amount of venom ejected. A rattlesnake, for example, may release much or little of its available venom into a bite. If it bites several times it can inject venom with each bite.

In rattlesnakes and moccasins (viperids), the fangs are long modified front teeth that normally are held against the roof of the mouth. In their bite position, the bone to which they are attached is pivoted down so they are roughly perpendicular to the roof of the mouth. Each fang is hollow, like a hypodermic needle, and is connected to the duct from the poison gland located on that side of the head. Snakes can control the position of their fangs in striking or in biting. In elapids—the coral snakes—the fangs are short and fixed in an erect position at the front of the mouth. In both viperids and elapids, the fangs are shed periodically and replaced by reserve fangs.

Venoms are complex mixtures of proteins, many having enzyme action. They cannot be categorized simply as hemotoxins, neurotoxins, or cardiotoxins, as was often done in the past. Most venoms have elements with each of these properties. The venoms of a species of snake may differ in different areas and at different times of the year. The amount of venom injected also varies. Field determinations of the severity of a bite are precarious. For these reasons, it is important to get the snakebite victim to a medical facility as quickly as possible.

In general, the venom of rattlesnakes and other vipers of North America causes local swelling through destruction of the tissues exposed to it, including blood vessels and blood cells. Swelling is usually evident within 10 minutes and may increase and spread slowly for several days, with much edema and discoloration in the bite area. Pain may be intense.

Numbness around the mouth, weakness, sweating, and muscle twitches are also characteristic following some rattlesnake bites.

In contrast, the venom of coral snakes is primarily neurotoxic. The effect on the tissue in the bite area is minimal, but the venom may cause heart or respiratory failure. Although more lethal, drop for drop, than viperid venom, few deaths are now attributed to North American coral snakes. The symptoms and signs of poisoning, including pain, may be delayed for more than half an hour after the bite, and there may be little or no swelling. Within two hours the victim may become drowsy, have blurred vision, and find it difficult to speak distinctly. A collapse of the respiratory mechanism may follow, necessitating artificial respiration. Get the victim to a hospital quickly.

First Aid

If a doctor or a hospital are not within 30 minutes distance, first aid must be administered either by the victim or a companion. Remain calm. Panic, exertion, and hysteria only help to spread the venom more rapidly. Remember, snakebite is rarely fatal.

Anyone who is collecting snakes or who is traveling where poisonous snakes occur should carry a snakebite first-aid kit. If one is not available, at-hand resources may be used.

First, at once place a constriction band immediately proximal to the bite. Then make a single, shallow, longitudinal (*not* cross) slit ⅛ to ¼ inch (3 to 6 mm) in length and depth over each fang wound, using a razor, knife blade, or other sharp object. Apply suction, preferably with a suction cup from a snakebite kit, but with the mouth, if necessary, if it has no lesions. These procedures must be started within 5 minutes of the bite to be effective, and the quicker the better. Suction should be carried out continuously for 30-60 minutes.

The longitudinal slit minimizes the chances of cutting vessels, nerves, tendons, and muscle fibers under the skin. A shallow cut is best because the venom generally at first pools in subcutaneous lymphatic channels. Rarely is the venom injected directly into a blood vessel, which would be followed almost immediately by alarming systemic signs and symptoms, or injected deep enough to pool between muscles or around bones where it would not be accessible to suction. The suction cups may need to be reapplied from time to time. The constriction band should be advanced ahead of the swelling.

If done well and promptly, this "cut and suck" first-aid treatment is of value. However, if the victim is within 30 minutes of medical care—rest, reassurance, immobilization of the injured part, and observation are all that is absolutely necessary. There is no harm in placing a constriction band just proximal to the wound site. However, a constriction band should only be tight enough to impede lymph flow. If swelling reaches the constriction band, move it farther away. When antivenin is given, the constriction band should be removed completely.

The victim should be kept as quiet as possible and should be kept warm. Shock may occur. Whiskey or other alcoholic drinks should *not* be given. Because some rattlesnake venoms cause changes in feeling in the throat, give water cautiously. Again, reassurance that everything will be all right is important.

Move the victim to a doctor or to a hospital where medical attention can be administered as needed. The victim should not walk if it can be avoided. It is best for the victim to lie quietly in a warm but airy place. Be sure to tell the doctor or medical attendants what kind of snake bit the victim. If you are not sure of its identification, the snake should be killed and taken with you in a sack or on a stick. Explain all you have done and any reactions of the victim to the bite or your first-aid efforts.

FIVE TYPES OF POISONOUS SNAKES IN NORTH AMERICA

CORAL SNAKE

RATTLESNAKE

COPPERHEAD

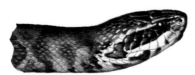

COTTONMOUTH

PIGMY RATTLESNAKE

11

REPTILE NAMES

Common names for reptiles are not yet fully standardized. Those used in this book conform to *Standard Common and Current Scientific Names for North American Amphibians and Reptiles.*

In this book, each species is designated first by a common name and then, in parentheses, by its scientific name. The scientific name, which has worldwide acceptance and documentation, generally also reveals the relationships of species. Subspecies are also designated.

Species and Subspecies

A *species* consists of genetically distinctive populations of individuals isolated reproductively from all other populations. A *subspecies* is a population segment of a species and occupies a particular portion of the range of that species. More than one subspecies of a species never occurs in the same locality (unless as end populations of a species with a circular range, which is rare). Species and subspecies are natural entities. All other divisions in the hierarchy of classification are conceptual—"man made" groupings that indicate degrees of similarity within or between species.

Scientific names for species consist of two words—first the generic name (genus) and then the specific name (species). *Thamnophis sirtalis* is thus the scientific name for a species known by the common name of Common Garter Snake. Note that the generic name always begins with a capital letter; the specific name does not. Both are italicized.

Subspecies bear a three-word scientific name. Twelve subspecies are recognized for the Common Garter Snake—six in eastern and six in western North America. One eastern subspecies is designated *Thamnophis sirtalis parietalis*. It differs from the basic or type (nominal) subspecies in having barred lips and red between its broad stripes. *Thamnophis sirtalis infernalis* is a similar western subspecies with black flecks rather than bars on its lips.

Higher Category Names

Above the level of species, the scientific name consists of a single word. The Common Garter Snake belongs to the large genus *Thamnophis*, which also contains the Mexican Garter Snake, the Plains Garter Snake, and others, plus the ribbon snakes. All members of the genus *Thamnophis* belong to the subfamily Natricinae in the family Colubridae. Names above the genus level are not italicized. Family names end in -idae and are based on the name of a member genus—in this case, *Coluber*, the genus containing racers. *Coluber* snakes, however, belong to the subfamily Colubrinae. Note that subfamily names end in -inae.

Basic classification categories for animals are, in descending order:

kingdom, phylum, class, order, family, genus, and species. Each may contain subdivisions, such as the subfamilies cited above.

All reptiles belong to the class Reptilia. In North America, they are represented by turtles (order Testudines), scaled reptiles (order Squamata, which is further subdivided into the suborder Lacertilia, or Sauria, containing the lizards; the suborder Amphisbaenia, containing amphisbaenids; and the suborder Serpentes, or Ophidia, containing snakes), and the crocodilians (order Crocodylia).

Kinds of Variations

Varied physical features occur at every classification level. This is called *diversity*. The kinds of variation include the following.

Clinal variation is common in species inhabiting a wide climatic range—either vertical as on mountain slopes or horizontal as on extensive plains. These species often are different at the extremities of their range, but the differences blend gradually from one extreme to another and are not uniform over extensive geographical areas.

Intergradation occurs where ranges of subspecies adjoin. It is a blending of the characters of two (or more) subspecies not completely separated geographically. Typically, the zone of intergradation between subspecies is narrow because each has adapted to different conditions. In some cases, however, the zone of intergradation is wide, and it is possible to find individuals with characteristics of two or more subspecies in the same population.

Hybridization refers to the production of offspring by parents of different species. Under some conditions it can occur between individuals of different genera or even different families. Hybrids are rare in nature, and they are usually sterile. In a few cases, hybridization has caused the formation of new species; nine whiptails in the United States originated from hybridization.

Age variations can make identification difficult, though less so than in amphibians that go through a complete body-form metamorphosis. Notable variations in the color and markings of juvenile reptiles are noted in the text in this book. Skinks are especially variable from juvenile to adult stages. The blue (or in a few cases, red) tail is believed to prevent aggression on the juveniles by adults of the same species.

Sex variations are most obvious during the mating period. Mature males and females otherwise tend to look alike. If differences are pertinent to making an identification, they are given. In a few lizard species—notably the collared and leopard lizards—the female develops orange blotches after mating.

Other notable variations are *albinism*, which refers to individuals lacking melanin (the dark pigment in the skin)—rare in reptiles; and *melanism*, which refers to an excess of melanin. Both conditions may be partial. *Morphs* are sharply distinctive color or pattern types occurring in some species.

13

HOW TO USE THIS BOOK

This guide focuses on the identification of adult reptiles. Concise information and precise illustrations are presented to enable you to identify all of the reptiles in North America north of Mexico. Field identification is emphasized, the book's material organized to speed and simplify identification. To take full advantage of this streamlined scheme, it is suggested that you do the following:

1. Scan the illustrations to find the one matching most nearly the specimen you have. This may result in an immediate identification.

2. If no illustration yields a positive identification, go through the illustrations again to find the group (family, genus, or species) with which the specimen you have shares one or more physical characteristics, such as color, markings, or anatomical features.

3. Inspect the range maps for that group. The maps tell which species occur where you found your specimen. They are color keyed to subspecies. (For color key, see the opposite page.)

4. Consult the species descriptions next to the maps. This narrows your choices until you arrive by elimination at the most likely identification—or perhaps more than one.

5. Verify your identification, or select among possible identifications, by using the keys.

Identifying by Keys

Keys are shortcuts to positive identification. They work by utilizing the presence or absence of "key" characteristics to eliminate one possible species after another until only the valid choice remains. Keys are not infallible, but they rank among the most important tools of the professionals. When easy to use, like the condensed, illustrated keys in this book, they can be valuable to the amateur as well. With turtles, as an example, you can begin with the key to the major groups (p. 22). This will lead you to the page on which you will find a second key that will provide the identification. In a few cases, the second key may direct you to a third or final key.

Suppose your specimen has stumplike or elephantine hind legs. You can tell from the very first key that you have a tortoise. When you turn to the page indicated, you will see that it belongs to the family Testudinidae and that only three species are native to the region covered by this book. In this case, the maps alone are all you need for the final identification if you know where the tortoise was obtained. Otherwise, the species descriptions will give the differences.

If the first key leads you to a second key, you simply work through the characters to your goal. You start with the first paired items and select the most fitting statement. This will direct you either to the common name of the specimen or to a boldface number of the next couplet to be consulted. Continue until you arrive at the common name.

After you have identified the species, you can narrow the identification even further by checking the distinguishing characters of the subspecies described in the text. For some species, no subspecies are currently recognized. Others have a number of subspecies. The range maps are especially helpful in identifying subspecies, for the subspecies of a species almost never occupy the same locality.

Policies of the Book

1. The sequence followed in presenting families and genera reflects their evolutionary development, going from the least advanced to the most advanced. Arrangements of species within those groups, however, is governed by considerations of clarity and space limitations.

2. To avoid complexity, only the most salient characters of subspecies are noted. Some herpetologists give each subspecies a common name. In this book, subspecies are treated as secondary in rank to species—which indeed is the case—and so are not given common names. If the species is known, the subspecies usually can be identified by its locality. Therefore, the various common names ascribed to subspecies tend to confuse.

3. Illustrations depict adults unless otherwise noted. Where helpful, subspecies are illustrated. Supplementary illustrations show significant variations or details and, in a few instances, sexual differences. The male is designated by the symbol ♂, the female by ♀.

4. Dimensions are given in both inches and millimeters, with fractional inches rounded off to the millimeter equivalents. Sizes listed are the reported *maximums* for the species. (To measure turtles, see p. 16; lizards, p. 19; snakes, p. 20.)

5. The range of each species north of Mexico is shown on the range maps. Ranges of subspecies, indicated as portions of the range of the species, are designated by different colors. A key to the colors used on the maps is given below.

COLOR KEY FOR RANGE MAPS

The range of each species north of Mexico is shown by a color area (or areas) on the maps. When there are subspecies, different colors are used within the area to indicate which portion of the range each subspecies inhabits. Numbers in the following key refer to the numbers used in the text to designate each subspecies.

■ Subspecies (1).	■ Subspecies (4).	■ Subspecies (7).
■ Subspecies (2).	■ Subspecies (5).	■ Subspecies (8).
■ Subspecies (3).	■ Subspecies (6).	

TOPOGRAPHY

Compared with amphibians, reptiles are much more diverse in body form and size—from the shelled turtles to the legless snakes and almost dinosaurlike crocodilians, from diminutive Reef Geckos only slightly more than an inch long to gigantic Leatherback sea turtles measuring 8 feet (244 cm) long and weighing as much as 1,600 pounds (730 kg) and to American Alligators known to reach a length of more than 19 feet (579 cm).

Turtles

Shells make turtles distinct from all other reptiles and unique among vertebrates. This protective covering encloses even their limb girdles, which are inside the bony rib cage that forms the framework for the upper shell, or carapace. The lower shell, or plastron, is joined to the upper at the sides by a bony "bridge" that varies in size and width in different kinds of turtles.

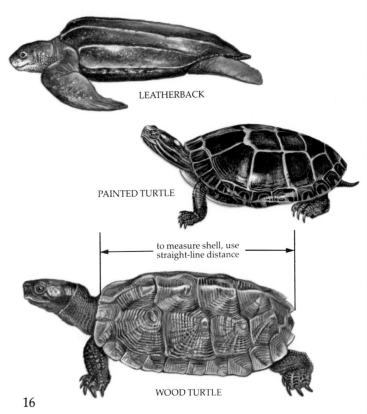

LEATHERBACK

PAINTED TURTLE

to measure shell, use straight-line distance

WOOD TURTLE

Terrestrial turtles have complete and high-domed shells. In box turtles, the plastron is hinged, enabling the turtles to draw in their head, limbs, and tail and then to shut the shell tightly. Aquatic turtles have a low-arched carapace and, in snapping turtles, the bridge is very short and narrow. The carapace has a leathery covering in softshell turtles and is largely leathery in the Leatherback sea turtle, which is distinguished also by the fatty ridges that run lengthwise on its back.

SPINY SOFTSHELL

GREEN TURTLE

growth rings

TEXAS TORTOISE

Unlike other reptiles, turtles do not have teeth, but their jaws do have sharp cutting edges. In the wholly aquatic marine turtles, both the front and hind limbs are modified as flippers for swimming. Freshwater turtles have distinct feet and clawed toes; their feet are broad, with webs between their widespread toes. Tortoises, the most terrestrial of the turtles, have modified limbs; their front legs are shovellike for digging or scraping and their hind legs are thick and elephantine.

Scaled Reptiles

Lizards, amphisbaenids, and snakes comprise this order of reptiles. Because of their obvious and distinctive external features, some authorities give each an order rank.

In North America, nearly all lizards have external ear openings, movable eyelids, and limbs. Exceptions are the glass lizards and legless lizards, which lack limbs but have movable eyelids and external ear openings; true and dwarf geckos and night lizards, which do not have movable eyelids but have limbs and ear openings; and the earless lizards, which have limbs and movable eyelids but no external ear openings. Lizards typically have five clawed toes on each foot and many rows of scales on their belly. Most North American lizards are small, measuring six inches (150 mm) or less in snout-to-vent length. Only a few species exceed 12 inches (300 mm) in length.

HEAD SCALES (as in *Sceloporus*)

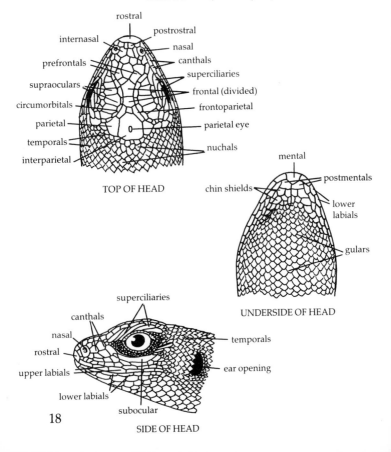

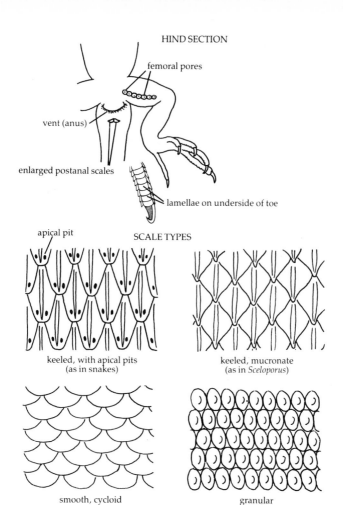

HIND SECTION

femoral pores

vent (anus)

enlarged postanal scales

lamellae on underside of toe

SCALE TYPES

apical pit

keeled, with apical pits
(as in snakes)

keeled, mucronate
(as in *Sceloporus*)

smooth, cycloid

granular

HOW TO MEASURE

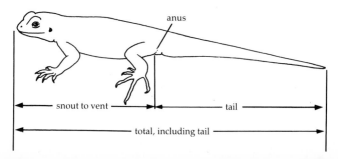

anus

snout to vent — tail

total, including tail

Snakes, as distinguished from lizards, have no external evidence of limbs, lack movable eyelids, and do not have external ear openings. Most snakes have a single row of broad, flat scales on the belly. These are attached to muscles and thus become aids in locomotion. The scales on the back may be either smooth or keeled (that is, with a ridge down the middle). As a group, snakes in North America are mostly less than three feet (912 mm) long, but several species do exceed six feet (1.8 m) in length.

HEAD SCALES

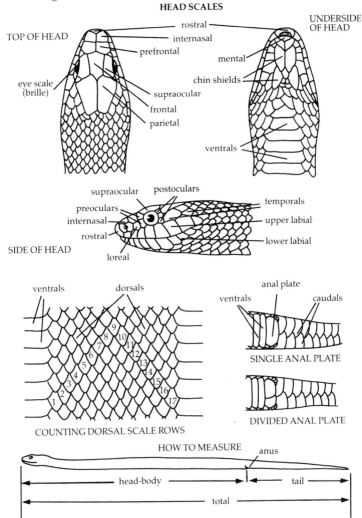

TOP OF HEAD

rostral
internasal
prefrontal
eye scale (brille)
supraocular
frontal
parietal

UNDERSIDE OF HEAD

mental
chin shields
ventrals

SIDE OF HEAD

supraocular
preoculars
internasal
rostral
loreal
postoculars
temporals
upper labial
lower labial

ventrals   dorsals

COUNTING DORSAL SCALE ROWS

anal plate
ventrals   caudals

SINGLE ANAL PLATE

DIVIDED ANAL PLATE

HOW TO MEASURE   anus

head-body   tail

total

20

Most aberrant of the scaled reptiles is the single amphisbaenid, the Florida Worm Lizard, which has a ringed body resembling an earthworm's, its scales small, flat, and rectangular. The tail is very short, and there is no external evidence of eyes or ears.

Crocodilians

Crocodilians, with two native and one introduced species in the United States, are unmistakable both because of size and appearance. They are semiaquatic, and have four limbs and a generally lizardlike body. The large head is projected into a snout rimmed with conical teeth. Both the eyes and nostrils protrude, enabling the animals to see and to breathe with only these "bumps" exposed above the water's surface. The skin is thick and leathery, covering bony scutes (or plates) on the back and smaller scales on the sides and belly. The five toes on each foot end in claws, except for the vestigial fifth toe of the hind foot. The large, long, laterally compressed tail supplies the power for swimming.

hatchling

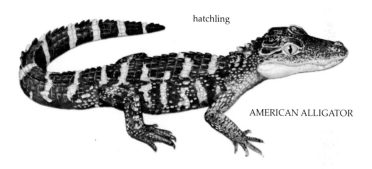

AMERICAN ALLIGATOR

GLOSSARY OF TERMS

Anterior: in front, or toward the head. **Caudal:** pertaining to or toward the tail. **Dorsal:** on the upper surface or back. **Dorsolateral:** on the upper side(s). **Lateral:** pertaining to the side(s). **Longitudinal:** running lengthwise. **Median:** in a middle or intermediate position. **Posterior:** behind, or toward the tail. **Ventral:** on the lower surface or belly.

TURTLES—order Testudines

All turtles have shells—an upper (carapace) and a lower (plastron), connected at the sides by bony "bridges." The carapace is formed over and fused to the ribs, and, most amazingly, the limbs and their girdles are located inside the ribs and shell. The steps in the evolution of this unusual arrangement have not yet been revealed by the fossil record. In some species, most notably the box turtles, the head and legs can be drawn completely inside the shell, which is hinged and can be closed tightly. In others the shell can be closed only partly or not at all.

All turtles lay eggs. The eggs may be flexible or hard-shelled, ovoid or spherical, and there may be from one to several hundred in each clutch. The females of some species lay two or more clutches each season. Some—the sea turtles, for example—do not lay eggs every year.

Whatever their normal habits or habitat, the female turtles must find suitable damp soil in which to dig shallow nests to hold their eggs. The nest is typically excavated with the hind legs, which are also used to cover the eggs. The forelegs are usually held in a fixed position during the digging and the egg laying. After the eggs are laid and covered, the females leave the nest site. They do not incubate the eggs or care for the young.

It is not always easy for males to find females, but turtles have evolved a remarkable ability to store viable sperm in the female for as long as seven years.

Turtles are toothless, but their beaks have sharp cutting edges. Most turtles are omnivorous or are wholly herbivorous. All must breathe air. Two sets of abdominal muscles contract alternately for breathing. All except box turtles and tortoises go into a state of dormancy underwater in cold regions, some for five or six months. Completely buried in mud, they get oxygen from the water through the linings of the mouth, throat, and posterior end of the gut, and in softshell turtles, through the skin covering the shell and body.

KEY TO MAJOR GROUPS OF TURTLES

1. Bridge much shorter than broad . . . . **snapping turtles,** p. 38
   Bridge at least as long as broad . . . . *see* **2**
2. All limbs oarlike . **sea turtles,** p. 34
   Limbs not oarlike . . . . *see* **3**
3. No laminae (scales) on shell; edges of shell flexible . . . . **softshell turtles,** p. 30
   Laminae present; shell is hard throughout . . . . *see* **4**

4. 6 pairs of laminae on plastron, none vestigial . . . . *see* **5**
   Anterior pair of plastral laminae vestigial, fused, or absent . . . . . . **mud and musk turtles,** p. 24
5. Hind feet stumplike, elephantine . . . . **tortoises,** p. 60
   Hind feet elongate . . . . **subaquatic turtles,** p. 40

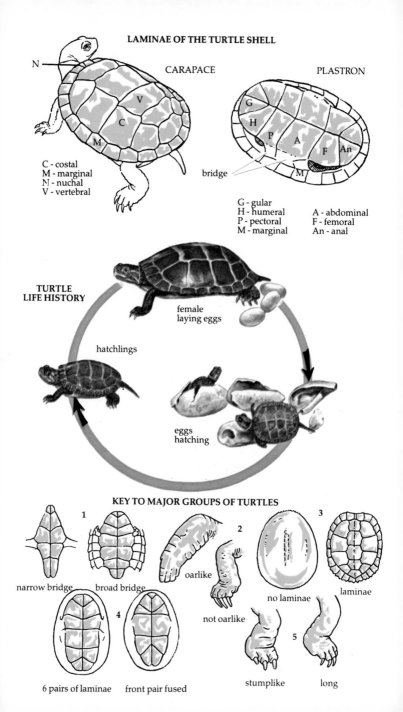

LAMINAE OF THE TURTLE SHELL

CARAPACE

PLASTRON

N

V
C
M

C - costal
M - marginal
N - nuchal
V - vertebral

G
H
P
A
F
An

bridge

M

G - gular
H - humeral
P - pectoral
M - marginal

A - abdominal
F - femoral
An - anal

TURTLE LIFE HISTORY

female laying eggs

hatchlings

eggs hatching

KEY TO MAJOR GROUPS OF TURTLES

1

narrow bridge   broad bridge

2

oarlike

not oarlike

no laminae

3

laminae

4

6 pairs of laminae   front pair fused

5

stumplike   long

MUD AND MUSK TURTLES—family Kinosternidae

This small family of bottom-dwelling aquatic turtles (four genera, about 21 species) ranges from Canada to northern South America. All species produce a strong odor from pairs of glands opening in the skin at the bridge. The plastron tends to be small. The carapace has 23 scutes (11 on each side, plus a nuchal) around the margin. In U.S. species, all plastral laminae (except gular) are paired. Males have a large, long tail ending in a blunt spine, an enlarged head, a deep notch at rear edge of plastron, and anus posterior to the edge of the carapace. Females have a short tail, small head, and rear edge of plastron smooth or only slightly notched. Both sexes have soft chin barbels. The forelegs are commonly used to dig nest; the clutch of 1-9 eggs is poorly concealed.

KEY TO MUD AND MUSK TURTLES

1. Greatest width of rear lobe of plastron more than half width of carapace at same level . . . . **mud turtles**, *see 5*
   Greatest width less than half width of carapace at same level . . . . **musk turtles**, *see 2*
2. Usually 2 light stripes on head; barbels on throat and chin . . . . **Stinkpot**, p. 28
   No light stripes; barbels on chin only . . . . *see 3*
3. Single gular lamina . . . . *see 4*
   No gular lamina (or vestigial) . . . . . . **Razorback Musk**, p. 28
4. *(not shown)* Shell more than twice as wide as high . . . .**Flattened Musk**, p. 28
   Not so . . **Loggerhead Musk**, p. 28

5. *(not shown)* 3 longitudinal light lines (may be dim or interrupted) on carapace . . **Striped Mud**, below
   No light lines . . . . *see 6*
6. 3rd marginal from rear center of carapace higher than 4th . . . . **Yellow Mud**, p. 26
   3rd marginal about same height as 4th . . . . *see 7*
7. Median anterior center lamina usually separated from 2nd paired marginal . . .**Eastern Mud**, p. 26
   Contacting 2nd paired marginal . . . . *see 8*
8. *(not shown)* Head shield notched at rear . . . . . . **Mexican Mud**, p. 26
   Not notched . **Sonoran Mud**, p. 26

MUD TURTLES—genus *Kinosternon*

Mud turtles have a less obnoxious odor than musk turtles. All of the skin over the plastron is cornified (horny) and, in adults, both ends of the plastron are hinged at the center. Gular lamina is well developed, pectoral laminae narrowly in contact. About 18 species—5 in U.S.

STRIPED MUD TURTLE (*Kinosternon bauri*). Smooth, oval shell. Males have patch of clasping tubercles on shank and thigh at contacting surfaces. Carapace has 3 longitudinal light lines, sometimes dim, obscured, or broken. Hatchlings have median keel, numerous longitudinal ridges; large light spot on underside of each marginal. A scavenger, often prowls out of water.

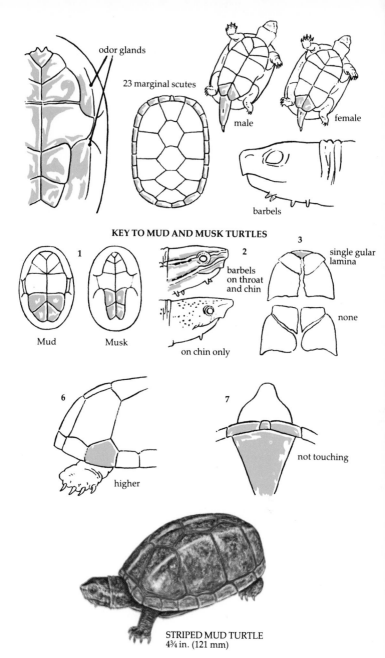

odor glands

23 marginal scutes

male

female

barbels

KEY TO MUD AND MUSK TURTLES

1

Mud   Musk

2

barbels on throat and chin

on chin only

3

single gular lamina

6

higher

7

not touching

STRIPED MUD TURTLE
4¾ in. (121 mm)

YELLOW MUD TURTLE (*Kinosternon flavescens*). Broad, flat shell. Head shield in adults lacks posterior notch; 9th and 10th marginal laminae higher than preceding marginals in those with shells exceeding 2.8 in. (70mm). Males have clasping tubercles. No stripes present on head. Young have low median keel, no lateral keels, all marginals about same height. Found chiefly in ponds and small streams with muddy bottoms, grasslands, or open woods. Three subspecies in U.S.: (1) *K. f. flavescens*—chin and neck bright yellow below and on sides, shell and head olive above, gular lamina less than half the length of front lobe of plastron; (2) *K. f. spooneri*—neck dark below, yellow only on chin, shell and head dark brown above; (3) *K. f. arizonense*—like (1) but gular lamina more than half the length of front lobe of plastron.

MEXICAN MUD TURTLE (*Kinosternon hirtipes*). Arched shell with low median keel (dorsolateral keels also present in very young). Males have clasping tubercles. Usually has 1 pair of chin tubercles; if more, posterior ones are smaller or located in front of middle of tympanum. A stream species. One subspecies in U.S.: *K. h. murrayi*.

SONORAN MUD TURTLE (*Kinosternon sonoriense*). Fairly flat shell with a low median keel, no dorsolateral keels (except in very young). Males have clasping tubercles. There are always 2 pairs of chin tubercles, the posterior pair level with middle of tympanum. Throat is mottled, with darker dorsal and lighter ventral tones broadly blended.

EASTERN MUD TURTLE (*Kinosternon subrubrum*). Smooth shell with a low median keel (broken dorsolateral keels in young). Males have poorly developed or no clasping tubercles. Carapace is yellowish to black (light marginal spots in young). Plastron is yellowish to brownish (in young, whitish to orange or orange-red, with dark in center and along seams). Head is spotted, mottled, or streaked. Other soft parts are olive-brown. Found in shallow, often brackish, water. Three subspecies: (1) *K. s. subrubrum*—no lines on head, bridge at least half the length of front lobe on plastron; (2) *K. s. hippocrepis*—2 continuous light lines on head; (3) *K. s. steindachneri*—no light lines on head, bridge less than half the length of front lobe on plastron, head enlarged in old males.

26

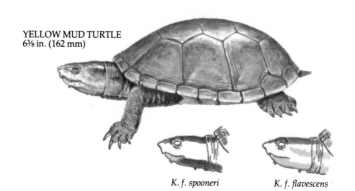

YELLOW MUD TURTLE
6⅜ in. (162 mm)

K. f. spooneri   *K. f. flavescens*

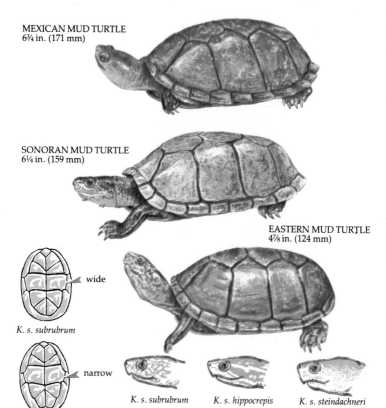

MEXICAN MUD TURTLE
6¾ in. (171 mm)

SONORAN MUD TURTLE
6¼ in. (159 mm)

EASTERN MUD TURTLE
4⅞ in. (124 mm)

wide

K. s. subrubrum

narrow

K. s. steindachneri

K. s. subrubrum   *K. s. hippocrepis*   *K. s. steindachneri*

MUSK TURTLES—genus *Sternotherus*

All musk turtles, which are limited to the eastern United States, have a strong odor—probably the strongest of all the turtles. The skin covering the central areas of the plastron never becomes cornified (horny), which leaves extensive patches of whitish skin that looks like cartilage. The plastron is distinctly smaller than in mud turtles, and only the anterior lobe is hinged. The gular lamina is absent or small, rarely paired, and the pectoral laminae are in broad contact along the midline. Most other characteristics are the same as those of mud turtles.

RAZORBACK MUSK TURTLE (*Sternotherus carinatus*). This turtle has a deep and somewhat compressed shell that is at least half as high as it is wide. The median ridge is distinct throughout the turtle's life. Males do not have clasping tubercles. The head is mottled but is not striped. This is a stream and swamp dwelling species.

LOGGERHEAD MUSK TURTLE (*Sternotherus minor*). Has 3 keels on the carapace, the 2 laterals less well developed than the median. Shell not twice as wide as high. A gular lamina is usually present and well developed. Males lack clasping tubercles. Dark brown above, pink to yellow below, unmarked. There are no light stripes on the head. This species is found in mud-bottomed, shallow water, both still and running. It climbs readily to bask, but basks only briefly. Two subspecies: (1) *S. m. minor*—dark dots on head and neck, head greatly enlarged in old males; (2) *S. m. peltifer*—dark stripes on head and neck, lateral keels sometimes absent.

FLATTENED MUSK TURTLE (*Sternotherus depressus*). Shell is more than twice as wide as high, 4½ in. (114 mm) long. Dark network on greenish head and neck. One gular lamina. Found in rocky streams.

STINKPOT (*Sternotherus odoratus*). May have a keeled, smooth, or even flattened shell in adults. The young sometimes have 3 keels. A gular lamina is usually present and is well developed. The males have a patch of clasping tubercles on the contacting surfaces of the shank and thigh. The sides of the head are black and usually have 2 conspicuous white or yellowish stripes. Found in streams, ditches, ponds, and lakes—almost any permanent body of water.

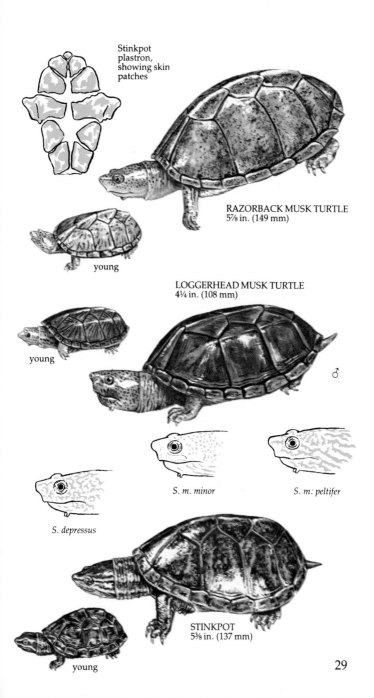

Stinkpot plastron, showing skin patches

RAZORBACK MUSK TURTLE
5⅞ in. (149 mm)

young

LOGGERHEAD MUSK TURTLE
4¼ in. (108 mm)

young

♂

S. depressus

S. m. minor

S. m. peltifer

STINKPOT
5⅜ in. (137 mm)

young

SOFTSHELL TURTLES—family Trionychidae

The seven genera and 22 species of this family occur in Africa, Asia, and North America. These turtles have a hard, bony shell under the skin—except at the edges, which are soft and flexible. The outline of the bones of the carapace and plastron can be seen beneath the leathery skin. There are no laminae on either the carapace or the plastron, which is reduced posteriorly. The body is flat and circular, and the snout is drawn out into tubular form, with the nostrils terminal. All four feet are fully webbed, broad, and flat, with three claws on each. In females, the claws are longer on the hind feet; in males, on the forefeet. In males, the tail is one-third the length of the carapace and protrudes far beyond the rear edge of the carapace; in females, the tail scarcely reaches the edge of the carapace. In both sexes, the anus opens near the tip of the tail rather than at the base, as in most turtles.

Softshells prefer waters that have a soft or muddy bottom and sandy banks. They often sun themselves along the edge, buried under a thin layer of sand in water that is shallow enough to permit them to breathe by stretching their long necks to the surface. Sometimes they bask out of the water, but are shy and wary, dashing for the water at the least disturbance. They are exceptionally fast swimmers and capable of catching fish, but their diet consists largely of small invertebrates. The Florida Softshell has broad crushing surfaces on its jaws, an adaptation for eating snails, clams, and other hard-shelled animals. Softshells are good to eat and are sometimes caught on hook and line; the Florida Softshell is common in markets.

As a rule, softshells have short tempers, making them difficult to handle as adults. With their long, flexible necks, they have an exceptional reach. Most females lay two to three clutches of eggs per season, each with 4–33 (average 15–20) spherical eggs that are 1.2 inch (30 mm) in diameter in the Spiny and Florida and 0.8 inch (20 mm) in diameter in the Smooth.

In cold regions, softshells go into a state of dormancy in mud underwater as do other freshwater turtles. After dormancy is started, they do not surface to breathe air. Gaseous exchanges through the skin and lining of the throat suffice for respiratory needs.

KEY TO SOFTSHELL TURTLES

1. Tip of snout flat; nostrils terminal and bean-shaped, each with ridge from median septum; in adults, carapace is tuberculate anteriorly . . . . *see* **2**
Tip of snout rounded; nostrils sublateral and oval, median septum without ridges; carapace never with tubercles . . . .
. . . . . **Smooth Softshell,** p. 32

2. Longitudinal series of tubercles on carapace; marginal ridge . . . . .
. . . . . . **Florida Softshell,** p. 32
Tubercles scattered; no marginal ridge . . . **Spiny Softshell,** p. 32

female

♀

male

♂

KEY TO SOFTSHELL TURTLES

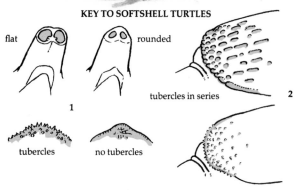

flat   rounded

tubercles in series

1

2

tubercles   no tubercles

tubercles scattered

FLORIDA SOFTSHELL (*Trionyx ferox*). Has a marginal ridge on the gray to brown carapace, with tubercles scattered over the anterior part of the carapace and clustered on anterior edge and marginal ridge. Tubercles are present on longitudinal ridges of juveniles. In the young, carapace is yellowish olive, sharply marked with a yellow-edged border and a coarse, light reticulum that separates irregular, large dark areas. Light markings become less evident with age. The plastron is gray to white and unmarked in both juveniles and adults. There is often a red or yellow stripe from rear of eye to angle of jaw, brightest in young.

SMOOTH SOFTSHELL (*Trionyx muticus*). Lacks tubercles or spines on the carapace, which is completely smooth, light-edged, and uniformly olive to orange-brown or with irregular darker markings. There is a black-edged light line through each eye. Plastron is unmarked white or gray. Some have dark flecks on the sides and upper surfaces of the neck and also on the front legs in young, which generally also have a pale yellow border around the sides and rear of the carapace. Two subspecies: (1) *T. m. muticus*—light stripe in front of the eye, juvenile pattern of dots and short lines; (2) *T. m. calvatus*—no light stripe in front of eye, juvenile pattern of large circular spots.

SPINY SOFTSHELL (*Trionyx spiniferus*). Has numerous tubercles or spines scattered (not on ridges or in lines) over the carapace, prominent in adults but scarcely evident in young. Two separate, dark-edged light stripes on each side of the head. Juveniles have pattern of ocelli, retained by older males, replaced by blotches in older females.

Six subspecies: (1) *T. s. spiniferus*—dark-edged ocelli (eye-sized or larger) or dark blotches and 1 dark marginal line on carapace; (2) *T. s. hartwegi*—young with smaller ocelli than in (1) on carapace, dark blotches in adults; (3) *T. s. asperus*—2 or more submarginal black lines paralleling rear margin of carapace; (4) *T. s. emoryi*—light border of carapace 4-5 times wider at rear than at sides, white tubercles on rear third of carapace; (5) *T. s. guadalupensis*—black-ringed white tubercles on most of carapace; (6) *T. s. pallidus*—white tubercles (no black rings) on posterior half of light carapace.

FLORIDA SOFTSHELL
female: 18 in. (457 mm)
male: 11 in. (279 mm)

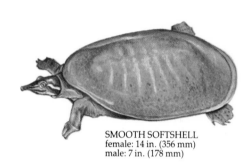

SMOOTH SOFTSHELL
female: 14 in. (356 mm)
male: 7 in. (178 mm)

SPINY SOFTSHELL
female: 20 in. (508 mm)
male: 9½ in. (241 mm)

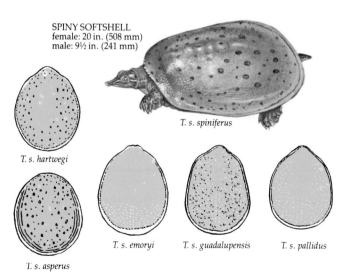

T. s. spiniferus

T. s. hartwegi

T. s. emoryi

T. s. guadalupensis

T. s. pallidus

T. s. asperus

SEA TURTLES—families Cheloniidae and Dermochelyidae

Sea turtles inhabit all temperate and tropical oceans; rarely wandering into colder zones. They are air breathers and must surface from time to time, even while sleeping, to replenish their oxygen. Their limbs are oarlike, the digits fused and identifiable only by the terminal claws. The tear duct is closed, so the excess secretions from the glands around the eyes flow out over the lower eyelids.

Sea turtles eat marine invertebrates and seaweeds. Green Turtle adults are the most herbivorous. Ridleys are the most exclusively carnivorous. The Leatherback eats principally jellyfish.

Nesting occurs on the sandy shores of warm oceans, the females coming from hundreds or even thousands of miles away to nest near where they hatched. Males accompany them to the offshore waters where they mate after the female lays her eggs. The Atlantic Ridley nests during the day and the Loggerhead usually does so; all others nest at night. The nest is dug above the high-tide level, to a depth of 1.5–2.5 feet (460–760 mm). The eggs (30–175) are spherical and 1.5–1.6 inches (38–40 mm) in diameter, except in the Green Turtle (1.8 in.; 46 mm) and the Leatherback (2.2 in.; 56 mm). The female returns one to five times, at intervals of 10 days to six weeks, to lay other batches of eggs. After an egg-laying season, she may not return for several years. Ridleys have the shortest cycles (one to two years); most others are two to three years. Some lie dormant in winter months in temperate zones, on the ocean bottom in shallow seas.

The Leatherback is placed in a separate family—Dermochelyidae—and sometimes even in a different suborder because of its unique leathery shell. It is the only strictly pelagic sea turtle. The others follow the continental shelves except when moving to breeding or feeding territories.

The survival of sea turtles is severely threatened primarily because of their susceptibility to predation while nesting, incubating, and hatching. Adults are sought as food, the Hawksbill for "tortoise shell." Only the Leatherback is considered inedible by humans, although all sea turtles can be fatally poisonous if they have been feeding on deadly seaweeds, which they are immune to.

KEY TO SEA TURTLES

1. Shell leathery. **Leatherback,** p. 36 Shell bony, with horny laminae. *see* 2
2. 4 lateral laminae on each side *see* 3 5 or more lateral laminae . . . *see* 4
3. 1 pair of scutes in contact on top of head between eyes . . . .**Green,** p. 36 2 pairs . . . . **Hawksbill,** p. 36
4. Bridge with 3 poreless laminae in a row on each side . . . . **Loggerhead,** p. 36 4 pore-bearing laminae . . . . *see* 5
5. Usually 5 lateral laminae on each side, gray . . . . **Atlantic Ridley,** p. 36 Usually 6 or more lateral laminae, olive . . . . **Pacific Ridley,** p. 36

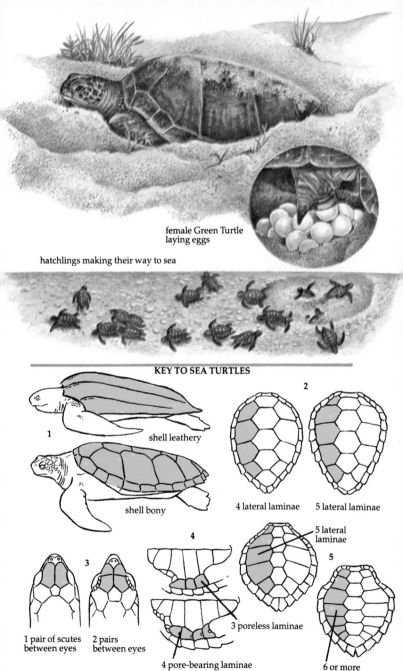

female Green Turtle
laying eggs

hatchlings making their way to sea

KEY TO SEA TURTLES

1
shell leathery

shell bony

2
4 lateral laminae   5 lateral laminae

3
1 pair of scutes
between eyes

2 pairs
between eyes

4
3 poreless laminae

4 pore-bearing laminae

5
5 lateral
laminae

6 or more
lateral laminae

LEATHERBACK (*Dermochelys coriacea*). Has 12 ridges on its leathery shell—5 above, 5 below, and 1 on each side. Possibly two subspecies: (1) *D. c. coriacea*—Atlantic Ocean; (2) *D. c. schlegeli*—Pacific and Indian oceans.

LOGGERHEAD (*Caretta caretta*). Has a bony shell, with 5 or more lateral laminae on each side of the carapace and 2 pairs of scutes in contact on the top of the head between the eyes. Reddish. Two subspecies: (1) *C. c. caretta*—Atlantic Ocean; (2) *C. c. tarapacana*—Pacific and Indian oceans.

GREEN TURTLE (*Chelonia mydas*). Has a bony shell, with 4 lateral laminae on each side of brownish carapace. This and the following species are the only sea turtles with just 4 lateral laminae on each side of the carapace. There are two subspecies: (1) *C. m. mydas*—Atlantic Ocean; (2) *C. m. agassizi*—Pacific and Indian oceans.

HAWKSBILL (*Eretmochelys imbricata*). Similar to the Green Turtle but has 2 pairs of scutes in contact on top of the head between the eyes. The head is large. Throughout most of its life, the Hawksbill has translucent, brown-flecked, and overlapping laminae on its carapace. The young have 2 prominent ridges on the plastron, and the laminae do not overlap on the carapace. In older individuals also the overlapping is sometimes lost. Two subspecies: (1) *E. i. imbricata*—Atlantic Ocean; (2) *E. i. bissa*—Pacific and Indian oceans.

ATLANTIC RIDLEY (*Lepidochelys kempi*). Has a 3-ridged carapace about as wide as it is long, 5 lateral laminae on each side, and 4 inframarginals, each with a pore near its rear margin. Has 2 pairs of scutes in contact on top of the head between the eyes. Carapace gray. The dorsolateral keels become inconspicuous in older individuals. It breeds in huge congresses in limited areas in northern Mexico.

PACIFIC RIDLEY (*Lepidochelys olivacea*). Much like the Atlantic Ridley but slightly larger. It has 6–8 lateral laminae on each side and a uniformly olive shell in adults.

LEATHERBACK
8 ft. (244 cm)
1,600 lbs. (726 kg)

LOGGERHEAD
7 ft. (213 cm)
1,200 lbs. (544 kg)

GREEN TURTLE
5 ft. (152 cm)
850 lbs. (385 kg)

HAWKSBILL
3 ft. (91 cm)
280 lbs. (127 kg)

ATLANTIC RIDLEY
29 in. (737 mm)
100 lbs. (45 kg)

PACIFIC RIDLEY
30 in. (762 mm)
120 lbs. (54 kg)

SNAPPING TURTLES—family Chelydridae

The two species in this family have a very small plastron connected to the carapace on each side by a narrow bridge at least twice as wide as it is long. The plastron consists of only 9 or 10 central laminae, the gular either paired or single. The abdominal laminae are displaced onto the bridge. The tail is relatively long, 0.6 to 1.2 times as long as the carapace, and is armed with a crest of bony plates. Alligator Snappers open the mouth under water and lure fish with a pink growth on the tongue.

COMMON SNAPPING TURTLE (*Chelydra serpentina*). Has 3 low keels on the carapace, prominent in the young but becoming obscure as the turtle matures. The plates on the midline of the tail are higher than they are long. There are no supramarginal laminae, and the eyes are clearly visible from above. Lives in quiet or slow-moving waters, preferably with a muddy bottom and ample vegetation. Largely nocturnal, foraging along bottom; older turtles are more sedentary. Diet about a third plants and the remainder mainly fish and aquatic invertebrates. Often basks at the surface, rarely on land. When first caught, secretes an unpleasant odor. The spherical eggs, 0.8–1.3 in. (20–32 mm) in diameter and 11–83 (usually 20–30) per clutch, are laid in moist soil at a depth of 4–7 in. (100–180 mm).

Two subspecies in U.S. (a third in Central and South America): (1) *C. s. serpentina*—upper neck tubercles wart-like; (2) *C. s. osceola*—upper neck tubercles elongate, pointed.

ALLIGATOR SNAPPING TURTLE (*Macroclemys temmincki*). Has 3 prominent keels on the carapace throughout life. Plates on midline of tail not as high as long, lower than in the Common Snapping Turtle. Has a row of 3–8 supramarginal laminae on each side of carapace between marginals and laterals. Eyes, on sides of head, are not visible from above. Usually found in deep waters of large streams and lakes, preferring those with mud bottoms. Largely nocturnal, foraging as it creeps along the bottom and usually working upstream. Never seems to bask but surfaces frequently to breathe. Food is almost entirely animal matter of all sorts, the tongue used as a lure mainly during the day when turtle is sedentary. The 15–30 spherical eggs, 1.2–1.9 in. (30–48 mm) in diameter, are laid on land near water in nests that are 6–14 in. (140–360 mm) deep.

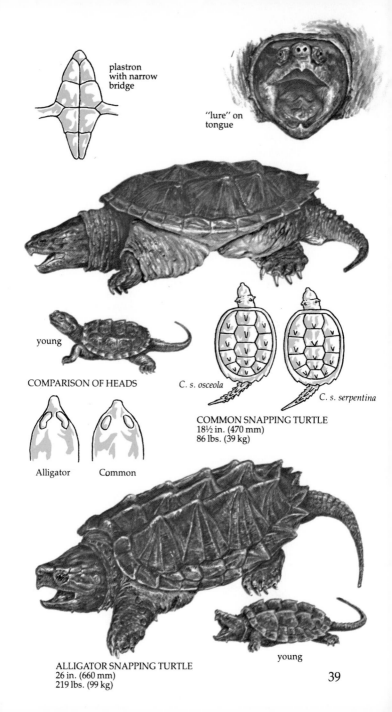

plastron with narrow bridge

"lure" on tongue

young

COMPARISON OF HEADS

Alligator   Common

C. s. osceola   C. s. serpentina

COMMON SNAPPING TURTLE
18½ in. (470 mm)
86 lbs. (39 kg)

ALLIGATOR SNAPPING TURTLE
26 in. (660 mm)
219 lbs. (99 kg)

young

39

SUBAQUATIC TURTLES—family Emydidae

Worldwide in distribution in tropical and temperate regions (except Australia and some islands), this group includes about 25 genera and 80 species, with 8 genera and 26 species occurring in the United States. They are divided into three distinct groups: (1) Chicken and Blanding's turtles, with a long neck, narrow jaw surfaces, and a preference for lotic (standing) waters; (2) diamondback terrapins, basking, and map turtles, with a short neck, broad jaw surfaces, and a preference for lentic (running) waters; and (3) pond and box turtles, with a short neck, narrow jaw surfaces, and a tendency to live on land.

Subaquatic turtles are all fully shelled, with laminae over the shell. The tail is short, and the bridge is broad. Some of the well-developed plastral laminae, always 12 in number (6 pairs), are in contact with the marginal laminae of the carapace at the bridge.

In contrast to tortoises, subaquatic turtles have hind feet that are more elongate and can be used for swimming as well as for walking on land.

Since most species of this family are aquatic, they are far more abundant both in individuals and in species in the humid Southeast than in the drier Midwest and West. Only southern Asia has a comparable abundance of species and genera, all different from those of America.

KEY TO SUBAQUATIC TURTLES

1. Blunt central beak on upper jaw; plastron hinged in adults, shell fully closable . . . . **box turtles,** p. 46
   Notch in center of upper jaw, or edges even; no beak; plastron not fully closable . . . . *see* **2**
2. Vertical bands on rear surface of thighs . . **Chicken Turtle,** p. 44
   No vertical bands . . . . *see* **3**
3. Deep groove or hinge between pectoral and abdominal laminae on plastron . . . . **Blanding's Turtle,** p. 44
   Groove between pectoral and abdominal laminae no deeper than elsewhere . . . . *see* **4**
4. Contact of abdominal laminae with marginals about equal to length of seam between abdominals . . . . **pond turtles,** p. 42

Contact no more than three-fourths the length of seam between abdominals . . . . *see* **5**
5. Growth rings clearly evident on laminae or carapace; neck speckled or streaked with dark on a light background . . . . **diamondback terrapins,** p. 60
   Growth rings scarcely or not evident; neck markings absent or consisting of light streaks or spots on dark background . *see* **6**
6. Some light lines on sides of head transverse or vertical in position, not all longitudinal . . . . **map turtles,** p. 48
   All light lines on sides of head longitudinal or diagonal . . . . **basking turtles,** p. 54

Subaquatic Turtle

Tortoise

KEY TO SUBAQUATIC TURTLES

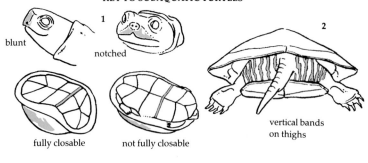

blunt

notched

1

fully closable

not fully closable

2

vertical bands on thighs

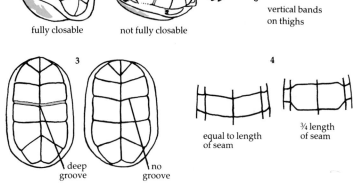

3

deep groove

no groove

4

equal to length of seam

¾ length of seam

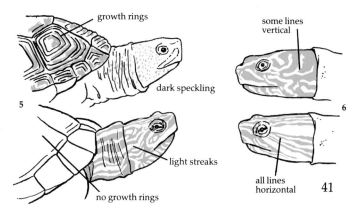

growth rings

dark speckling

some lines vertical

5

no growth rings

light streaks

6

all lines horizontal

41

POND TURTLES—genus *Clemmys*

Color and geographical distribution distinguish the four species in this genus. Spotted, Wood, and Bog turtles are equally terrestrial and aquatic. The Western Pond Turtle seldom leaves water except to bask. All are omnivorous, the Western the most carnivorous. The ovoid eggs, 1.2–1.6 inches (30–40 mm) in diameter, are laid in shallow, flask-shaped nests 2–2.5 inches (50–63 mm) deep, the neck 1.2 inches (30 mm) wide. Clutches contain 3–12 eggs. Males have a somewhat concave plastron. Eastern species are prized as pets; some are now rare owing to habitat destruction.

SPOTTED TURTLE (*Clemmys guttata*). Dark gray to black, usually with small round yellow spots scattered over the carapace and soft parts. Old individuals and hatchlings may be spotless. The plastron is extensively smudged with black or gray, otherwise yellow-orange in color. Males have a light brown throat and brown eyes; the throat is yellow and the eyes orange in females.

WOOD TURTLE (*Clemmys insculpta*). Shell is intricately sculptured with concentric growth rings and superimposed ridges radiating from central growth ring on each dorsal lamina. Carapace is gray to brown, the soft skin darker. Adults have a suffusion of orange or red on the soft skin, with indistinct dark and yellow lines sometimes visible radiating from growth centers on dorsal laminae. Plastron is yellow, each lamina with an irregular lateral blotch.

BOG TURTLE (*Clemmys muhlenbergi*). A small, brown-shelled turtle with an irregular, distinctive, large yellow to red blotch behind each eye. Growth rings evident on laminae but do not pyramid as in Wood Turtle. Plastron dark-smudged, with central lighter area.

WESTERN POND TURTLE (*Clemmys marmorata*). Has a lower, more streamlined shell than other pond turtles, and the yellowish plastron is less extensively dark-marked. Carapace is uniformly gray to brown or black, or has inconspicuous lines or rows of spots radiating from center of each lamina. Skin grayish with some yellowish suffusion. Two subspecies: (1) *C. m. marmorata*—head and neck reticulated with dark lines; (2) *C. m. pallida*—head and neck unmarked.

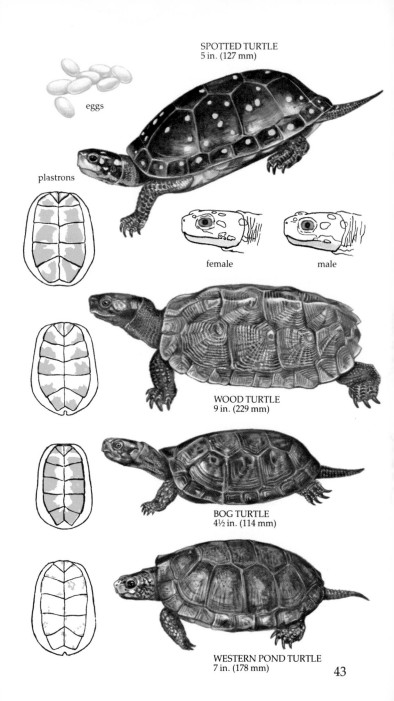

eggs

plastrons

SPOTTED TURTLE
5 in. (127 mm)

female

male

WOOD TURTLE
9 in. (229 mm)

BOG TURTLE
4½ in. (114 mm)

WESTERN POND TURTLE
7 in. (178 mm)

43

CHICKEN AND BLANDING'S TURTLES
—genera *Deirochelys* and *Emydoidea*

Although markedly different, these two species are closely related. Both have very long necks and narrow, ridgeless jaw surfaces.

CHICKEN TURTLE (*Deirochelys reticularia*). Inhabits ponds, lakes, swamps, and marshes, rarely if ever occurring in streams or rivers. Neck is three-fourths or more the length of the shell. Also distinctive is pattern of alternating light and dark vertical stripes on rear surface of thighs. A broad, light line follows exposed surface of foreleg. Carapace has no keel and no serrations at rear. Tip of the jaw has feeble notch. Carapace has a coarse network of light lines of varying distinctness. Females lay 5–15 eggs, 1.2×.8 in. (30×20 mm), in nests 4–5 in. (100–127 mm) deep. These are omnivorous turtles, but they have a preference for small invertebrates.

Three subspecies: (1) *D. r. reticularia*—plastron unmarked yellow, lines on carapace narrow and moderately distinct; (2) *D. r. chrysea*—plastron unmarked yellow, lines on carapace broad and conspicuous; (3) *D. r. miaria*—plastron yellow with dark marks along seams, lines on carapace broad but faint.

BLANDING'S TURTLE (*Emydoidea blandingi*). Has an exceptionally long neck, like Chicken Turtle. Carapace is speckled with yellow, the plastron hinged much as in box turtles. Not able to close shell completely, and has notch instead of hook at tip of upper jaw. Shell is not as highly arched as in box turtles. Lower jaw and throat are yellow, and there is a dark blotch at side of each lamina of plastron—so large in some that the whole plastron looks dark. Except in young, carapace lacks a keel. In young, carapace may be uniformly dark; it lacks a hinge and has a central dark figure.

Completely aquatic except for basking, lives mostly in still, shallow muddy water with vegetation, as in marshes, sloughs, ponds, lakes, and creeks. Food is largely animals (crayfish and insects predominating) but with plants a small but regular part of the diet. The 6–11 eggs (1.4×.8 in.; 36×20 mm) are laid in a nest 6–7 in. (150×178 mm) deep.

young

striped thighs

CHICKEN TURTLE
10 in. (254 mm)

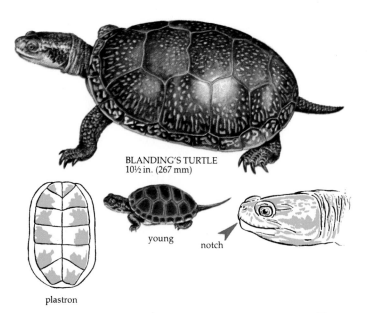

BLANDING'S TURTLE
10½ in. (267 mm)

plastron

young

notch

BOX TURTLES—genus *Terrapene*

These small terrestrial turtles have dome-shaped shells. Both the front and rear parts of the plastron operate on a single hinge, enabling the turtles to close the shell completely. In males, the plastron is at least slightly concave (strongly so in the Eastern) and claws of the hind feet are enlarged and curved as an aid in grasping the edge of the female's shell in copulation. The upper jaw has a median projection. Box turtles are strictly diurnal—the Eastern inhabiting primarily open woods, the Western prairies. Both are omnivorous, eating fruit, carrion, vegetables, flowers, insects, worms, and small vertebrates. The Eastern is chiefly herbivorous, the Western carnivorous. They are famed as pets, but the Western may bite. Box turtles go into a state of dormancy on land, a characteristic shared only with tortoises in North America. The ovoid eggs (2–8) measure 1.2 × .8 inches (30 × 20 mm) and are laid in a flask-shaped nest 2–3 inches (50 × 76 mm) deep and 3–4 inches (76 × 100 mm) wide that is dug in sandy soil.

EASTERN BOX TURTLE (*Terrapene carolina*). Adults have a plastron uniformly yellow or irregularly smudged with black, no radiating yellow lines (except sometimes in Florida), and hinge opposite 5th marginal. Carapace has a low but distinct keel. Hatchlings, seldom seen, are dark brown with light spot in center of each lamina above (except marginals), sometimes a median light stripe, a dark central figure on plastron, and no hinge. Four subspecies: (1) *T. c. carolina*—prominent yellow blotches on laminae of carapace, marginals slightly or not flared, usually 4 claws on hind feet; (2) *T. c. major*—larger than others, dorsal pattern obscure, marginals strongly flared, usually 4 claws on hind feet; (3) *T. c. triunguis*—no bright yellow markings on plastron, carapace obscurely marked or uniformly yellow, usually 3 claws on hind feet; (4) *T. c. bauri*—carapace and plastron with bright radiating lines, 2 stripes on each side of head, usually 3 claws on hind feet.

WESTERN BOX TURTLE (*Terrapene ornata*). Occasionally hybridizes with Eastern. Usually has no keel, always has radiating yellow lines on plastron, and plastral hinge opposite 6th marginal or between 5th and 6th. Hatchlings have median light stripe, numerous flecks of yellow on carapace, and no keel. Two subspecies: (1) *T. o. ornata*—darker, 5–8 lines on 2nd lateral lamina; (2) *T. o. luteola*—lighter, old adults often uniformly colored, 11–14 lines on 2nd lateral lamina.

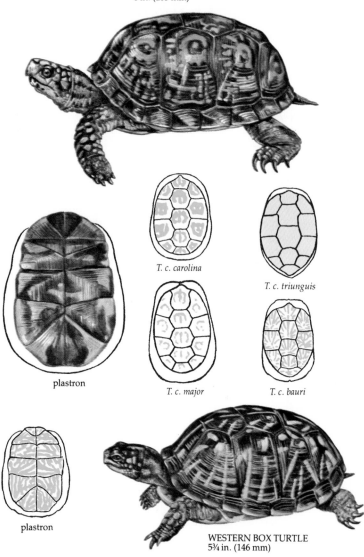

EASTERN BOX TURTLE
8 in. (203 mm)

plastron

T. c. carolina

T. c. triunguis

T. c. major

T. c. bauri

plastron

WESTERN BOX TURTLE
5¾ in. (146 mm)

MAP TURTLES—genus *Graptemys*

These highly aquatic turtles of ponds, lakes, streams, and rivers have broad crushing surfaces on their jaws, an adaptation for their diet of such hard-shelled creatures as snails, clams, crayfish, and insects. They will eat almost any kind of animal, however, and also consume carrion and some plants. Basking sites, such as rocks or fallen trees, are an essential. There are ten species, limited to the eastern U.S. and Canada.

Hatchlings of this group have a serrated keel on the carapace that is retained by males but may be greatly reduced or absent in females.

There is great sexual dimorphism in adult size, with females sometimes more than twice as long as males. In some, notably the Alabama and Barbour's Map, females have abnormally large heads for turtles. The difference in body and head size of males and females allows them to feed on different prey, reducing competition for food.

Males of all except the Map, Barbour's Map, and Alabama Map have long nails on their forefeet. In courtship, the male caresses the sides of the female's head with the backs of the nails. The flask-shaped nests are about 6 inches (150 mm) deep, with a 2–inch (50–mm) opening and an egg chamber 3 inches (76 mm) wide. Two or three clutches, each with 3–20 eggs, are laid per season. The eggs are ovoid, 1.2–1.8 inches (30–46 mm) long and .8–1.2 inches (20–30 mm) wide.

KEY TO MAP TURTLES

1. Height of projection at rear of 2nd central lamina one-sixth or more the length of the same lamina . . . . *see* **6**
Less than one-sixth the length . . . . *see* **2**

2. Light mark behind eye isolated . . . . **Map,** p. 50
Continuous with other marks . . . . *see* **3**

3. (*not shown*) Spot behind eye reddish . . . . **Texas,** p. 50
Spot yellowish . . . . *see* **4**

4. Spot behind eye large and C-shaped, transverse yellowish bar across lower jaw . **Cagle's,** p. 50
Not both of above . . . . *see* **5**

5. (*not shown*) Large yellowish spots beneath eye and on lower jaw . . . . **Ouachita,** p. 50
No large spots in these areas . . . .**False Map,** p. 50

6. Spot behind eye large, no other light lines reaching eye . . *see* **7**
Spot small, other light lines reaching eye . . . . *see* **8**

7. Transverse light bar behind chin; light marks on marginals narrow, less than half the width of marginals . . . . **Barbour's,** p. 52
Longitudinal light line behind chin; light marks on marginals broad, more than half the width of marginals . . **Alabama,** p. 52

8. Narrow, dark-edged light line forming a ring on each lateral lamina . . . . *see* **9**
Large central light area on each lateral lamina . . . . **Yellow-blotched,** p. 52

9. (*not shown*) Projections on central laminae blunt, rounded; mark behind eye continuous with neck stripe . . . **Black-knobbed,** p. 52
Projections narrow, compressed, pointed; mark behind eye usually not connected with neck stripe . . . . **Ringed,** p. 52

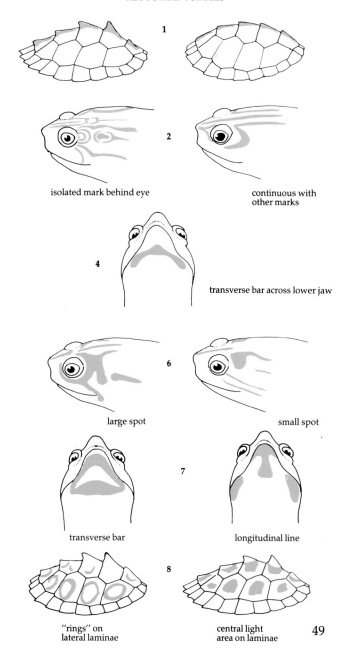

1

isolated mark behind eye

2

continuous with other marks

4

transverse bar across lower jaw

large spot

6

small spot

transverse bar

7

longitudinal line

8

"rings" on lateral laminae

central light area on laminae

CAGLE'S MAP TURTLE (*Graptemys caglei*). A small species from the Guadalupe River system, to which it is restricted. Greenish carapace with black or brown median projections and cream-colored lines or small blotches. Most individuals—88% of those checked—have the J-shaped marks behind the eyes joined in a **V** on top of the head. The transverse, cream-colored line across the jaw is unique. Largely insectivorous. Hatchlings emerge from Sept. through Nov.

MAP TURTLE (*Graptemys geographica*). This is the least modified species in the genus, its low carapace having little evidence of median projections. The mark behind the eye is distinctive because it is isolated. The intricate pattern of thin yellowish lines on the olive carapace may be faded in old adults.

FALSE MAP TURTLE (*Graptemys pseudogeographica*). Has a narrow yellowish crescent-shaped mark behind the eye that is continuous with the neck stripe, and a lengthwise yellowish line on the chin. The False Map Turtle lacks light spots below the eye and on the lower jaw. Two subspecies: (1) *G. p. pseudogeographica*—mark behind eye short, with 4–7 neck stripes reaching eye; (2) *G. p. kohni*—long crescent-shaped mark behind eye, other stripes do not reach the eye.

OUACHITA MAP TURTLE (*Graptemys ouachitensis*). Formerly considered a subspecies of the False Map Turtle. Females are 8½ in. (216 mm) long, males 4 in. (102 mm). May be distinguished by 2 large yellowish spots, 1 under the eye and 1 on the lower jaw. Two subspecies: (1) *G. o. ouachitensis*—mark behind eye square or rectangular, 1–3 neck stripes reach the eye; (2) *G. o. sabinensis*—oval or elongate mark behind eye, 5–9 neck stripes reach the eye, transverse yellowish lines on chin.

TEXAS MAP TURTLE (*Graptemys versa*). Not well known, diminutive, and isolated species that is closely related to the False Map Turtle. Formerly it was regarded only as a subspecies. The reddish lengthwise streak on the neck continuous with the mark behind the eye is distinctive, as are the markedly convex laminae on the carapace. Chin is patterned with orange or yellow dark-bordered blotches. This species is found only in the Colorado River system.

CAGLE'S MAP TURTLE
female: 6½ in. (165 mm)
male: 3⅝ in. (92 mm)

MAP TURTLE
female: 10⅝ in. (270 mm)
male: 6¼ in. (159 mm)

FALSE MAP TURTLE
female: 10 in. (254 mm)
male: 5¾ in. (146 mm)

G. p. pseudogeographica

G. p. kohni

young

G. o. ouachitensis

G. o. sabinensis

TEXAS MAP TURTLE
female: 4¾ in. (121 mm)
male: 3½ in. (89 mm)

BARBOUR'S MAP TURTLE (*Graptemys barbouri*). Projections on the central laminae are not evident in some old specimens but are very prominent in others. The large semicircular mark behind the eye is similar to that of the Mississippi Map and Alabama Map turtles. Distinctive features are narrow, light marks on the marginal laminae and a transverse light bar on the anterior part of the throat. In clear, limestone streams with fallen trees for basking sites. Very shy.

ALABAMA MAP TURTLE (*Graptemys pulchra*). Closely related to Barbour's Map Turtle. Older adults have low central projections on the carapace. Distinctive features are longitudinal rather than transverse chin stripe and broad, light bars on marginals. Limited to deep, slow streams over sand or gravel. Very shy.

RINGED MAP TURTLE (*Graptemys oculifera*). Has high, compressed, pointed central projections on carapace throughout its life. Usually isolated, comma-shaped mark behind eye is distinctive. Yellow or orange ocelli are prominent on laterals and marginals. Skin black. Likes swift streams over sand or clay.

BLACK-KNOBBED MAP TURTLE (*Graptemys nigrinoda*). This turtle has a broad, rounded, black, knoblike projection from the rear of each of its anterior 4 vertebral laminae. Lateral margin of carapace serrated. Mark behind orbit of eye is connected with longitudinal neck stripe united on midline with its mate from the opposite side. The ocelli on laterals and marginals are prominent, like those of the Ringed Map Turtle. Two subspecies: (1) *G. n. nigrinoda*—dark markings on a maximum of one-third of the plastron, skin mostly yellow; (2) *G. n. delticola*—dark markings over two-thirds or more of the plastron, skin mostly black.

YELLOW-BLOTCHED MAP TURTLE (*Graptemys flavimaculata*). Has high, compressed, pointed projection on anterior vertebral laminae. The mark behind orbit of the eye is subrectangular and connected with a dorsolateral light line that does not fuse with its mate. Other lateral neck lines reach the eye. There is a large central yellow blotch on each lateral and central lamina, and a wide longitudinal yellow bar or semicircle on each marginal. Skin olive.

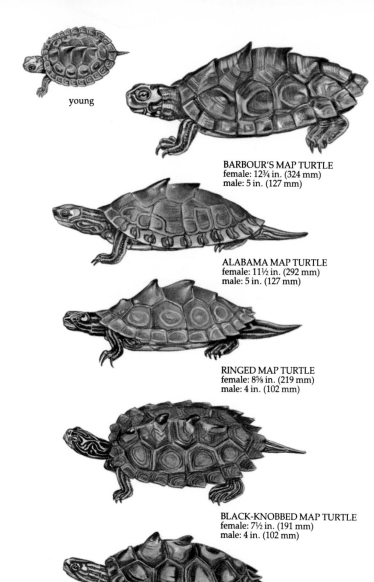

young

BARBOUR'S MAP TURTLE
female: 12¾ in. (324 mm)
male: 5 in. (127 mm)

ALABAMA MAP TURTLE
female: 11½ in. (292 mm)
male: 5 in. (127 mm)

RINGED MAP TURTLE
female: 8⅝ in. (219 mm)
male: 4 in. (102 mm)

BLACK-KNOBBED MAP TURTLE
female: 7½ in. (191 mm)
male: 4 in. (102 mm)

YELLOW-BLOTCHED MAP TURTLE
female: 6¾ in. (171 mm)
male: 4 in. (102 mm)

53

BASKING TURTLES—genera *Chrysemys* and *Pseudemys*

All of the seven species, long placed in the single genus *Chrysemys*, still adopted by some authorities, are strongly aquatic but bask regularly. They inhabit soft-bottomed streams, lakes or ponds, marshes, and swamps where there is ample vegetation. Some species have distinct preferences. The Alabama Red-bellied Turtle, for example, is limited largely to swamps and marshes, the Red-bellied Turtle to ponds.

Basking turtles are largely diurnal. In northern latitudes, they go into a state of dormancy underwater, not breathing for four to five months. Respiration during dormancy takes place through paired cloacal bursae. One or two clutches of 2–23 eggs are deposited in flask-shaped nests 4–6 inches (100–150 mm) deep. Eggs vary from 0.6–1.6 inches (15–40 mm) in length and 0.8–1.2 inches (20–30) in width, the smaller sizes in smaller species.

Diet is correlated with the width of the crushing surfaces of the jaws and with the presence or absence of a ridge and cusps. The Painted Turtle, which has the narrowest crushing surfaces and no ridge, is mainly carnivorous, though adults eat considerable amounts of plant food. The Slider, which has a cuspless ridge, is omnivorous, relying almost equally on plant and animal foods. All other species have cusped ridges and are almost exclusively herbivorous. The young tend to be carnivorous, shifting to plants as they mature. The animal food consumed is varied but is primarily insects and snails. Carrion is frequently eaten.

Mature males of all species are smaller than the females and have elongated claws on the forefeet. In courtship, the male swims to the front of the female and caresses the sides of her head with the backs of the claws on his outstretched forelegs.

KEY TO BASKING TURTLES

1. Apex of upper jaw with central notch flanked on each side by cusp . . . . *see* **2**
   No cusps flanking central notch, if present . . . . *see* **5**
2. Rear edge of carapace essentially smooth . . **Painted Turtle,** p. 56
   Rear edge notched . . . . *see* **3**
3. (*not shown*) Carapace with a reticulum of fine light lines . . . . *see* **5**
   Carapace mottled or with broad, light bar on each lamina . . *see* **4**
4. (*not shown*) 5 light lines on top of head between orbits . . . . *see* **7**
   3 light lines between orbits, paramedian lines terminating behind eye level . . . . **Florida Red-bellied,** p. 58

5. (*not shown*) Ridge on crushing surface of upper jaw without cusps . . . . **Slider,** p. 56
   With prominent cusps . . . . *see* **6**
6. Plastron unmarked except, at most, lines along seams . **Cooter,** p. 58
   Plastron with large central figure at least anteriorly . . . . **River Cooter,** p. 58
7. Carapace with fine pits and ridges . . **Alabama Red-bellied,** p. 58
   Carapace smooth . . . . **Red-bellied,** p. 58

basking in the sun

KEY TO BASKING TURTLES

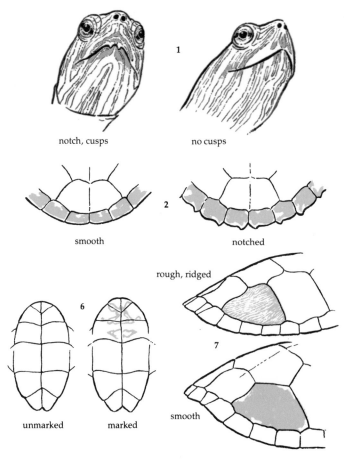

notch, cusps   1   no cusps

smooth   2   notched

rough, ridged

6

unmarked   marked

7

smooth

55

Four Species-groups of basking turtles are recognized: (1) Painted Turtle (*Chrysemys picta*), with narrow, flat crushing surfaces on jaws and a cusp on each side of a central notch in upper jaw; (2) Slider (*Pseudemys scripta*), with broader, ridged crushing surfaces on jaws and no cusp flanking the central notch in upper jaw; (3) Red-bellied Turtles (*P. alabamensis*, *P. nelsoni*, and *P. rubriventris*), with broad crushing surfaces bearing a cusped ridge on upper jaw, a cusp on each side of a central notch in upper jaw, and median line on head joining supratemporal lines at tip of snout to form a "rostral arrow"; and (4) Cooters (*P. concinna* and *P. floridana*), with a cusped ridge on crushing surfaces of upper jaw, no rostral arrow, and no cusp flanking a central notch in upper jaw (except in one subspecies). Identification in this genus has been complicated by release of specimens outside their normal range, where they may hybridize with native populations.

PAINTED TURTLE (*Chrysemys picta*). A brightly marked turtle, often sold in pet shops. Lines on head and neck do not vary greatly in width. Those on neck are red, becoming yellow on head. Carapace is low, without a keel except in hatchlings. Four subspecies: (1) *C. p. picta*—light border across aligned seams of rear lateral and central laminae, plastron unmarked; (2) *C. p. marginata*—small central figure on plastron; (3) *C. p. dorsalis*—median light line on carapace, plastron unmarked; (4) *C. p. belli*—fine network of light lines on carapace and large radiating central figure on plastron.

SLIDER (*Pseudemys scripta*). Has enlarged mark behind each eye—either widened line or isolated spot. Carapace has low keel and small notches on its rear edge. Four subspecies: (1) *P. s. scripta*—large yellow blotch behind each eye, continuous ventrally toward neck stripe, no pattern on at least middle and rear parts of plastron; (2) *P. s. troosti*—narrow yellow-orange stripe behind each eye, dark ocellus or smudge on each plastral lamina; (3) *P. s. elegans*—wide red stripe behind each eye and dark blotch on each plastral lamina (this is the Red-eared Slider that is commonly sold as a pet turtle); (4) *P. s. gaigeae* (considered a separate species by some authors)—isolated orange-red mark near rear of head (far back from eye), central figure on plastron or separate blotch on each plastral lamina, and carapace with reticulate pattern of light lines.

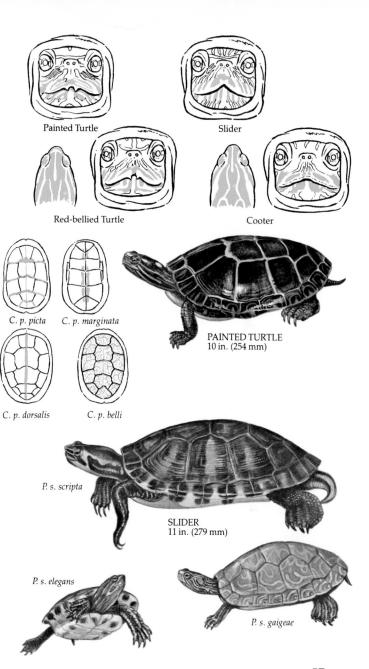

Painted Turtle

Slider

Red-bellied Turtle

Cooter

C. p. picta

C. p. marginata

C. p. dorsalis

C. p. belli

PAINTED TURTLE
10 in. (254 mm)

P. s. scripta

SLIDER
11 in. (279 mm)

P. s. elegans

P. s. gaigeae

57

RED-BELLIED TURTLE (*Pseudemys rubriventris*). Has reddish plastron and notch flanked by cusps at tip of upper jaw. Ridge on crushing surface of upper jaw bears several conical cusps. Light line on top of head joins transverse line bordering snout and forms "rostral arrow"; 5 light lines are present between eyes. Two subspecies (validity dubious): (1) *P. r. rubriventris*—carapace higher (42% of carapace length); (2) *P. r. bangsi*—nearly exterminated, relics in Plymouth Co. and on Naushon Island, Massachusetts, carapace lower (38% of carapace length).

ALABAMA RED-BELLIED TURTLE (*Pseudemys alabamensis*). Similar to Red-bellied but has rough carapace in adults, with fine ridges and pits.

FLORIDA RED-BELLIED TURTLE (*Pseudemys nelsoni*). Like Red-bellied but has only 3 light lines between eyes; paramedian lines end back of eye level.

RIVER COOTER (*Pseudemys concinna*). Has well-developed cusp-flanked notch in 1 subspecies, in occasional individuals of another (none in Cooter). Distinguished by strongly cusped ridge on crushing surfaces of upper jaw—from all except Cooter, with which it frequently hybridizes. Has patterned plastron, C-shaped light mark on 2nd lateral lamina, 2 bones in the 5th toe, single ocellus on underside of each marginal lamina. Prefers rivers. Five subspecies: (1) *P. c. concinna*—head-neck stripes yellow, stripes on outer surface of hind foot; (2) *P. c. suwanniensis*—head-neck stripes yellow, no stripes on outer surface of hind foot; (3) *P. c. hieroglyphica*—shell indented at bridge; (4) *P. c. mobilensis*—head-neck stripes orange-red; (5) *P. c. texana*—cusp-flanked median notch in upper jaw, stripe behind eye usually replaced by isolated yellow mark behind eye.

COOTER (*Pseudemys floridana*). Has no marks on plastron. Light line on 2nd lateral lamina is straight, forked at one or both ends. Has 3 bones in 5th toe, multiple ocelli on lower surfaces of marginal laminae. Prefers lakes and ponds. Three subspecies: (1) *P. f. floridana*—5 lines on top of head, not united to form loops on snout; (2) *P. f. peninsularis*—loops on snout; (3) *P. f. hoyi*—lines on top of head broken.

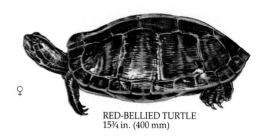

♀

RED-BELLIED TURTLE
15¾ in. (400 mm)

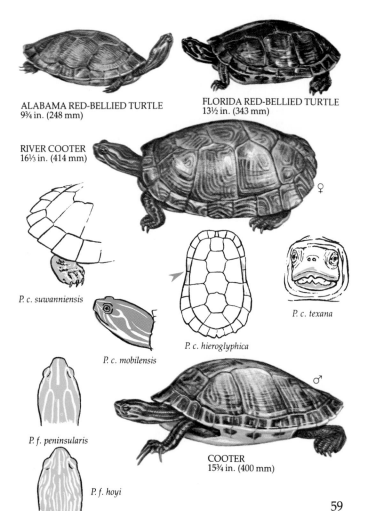

ALABAMA RED-BELLIED TURTLE
9¾ in. (248 mm)

FLORIDA RED-BELLIED TURTLE
13½ in. (343 mm)

RIVER COOTER
16⅓ in. (414 mm)

♀

P. c. suwanniensis

P. c. mobilensis

P. c. hieroglyphica

P. c. texana

P. f. peninsularis

P. f. hoyi

♂

COOTER
15¾ in. (400 mm)

DIAMONDBACK TERRAPINS—genus *Malaclemys*

Terrapins, limited to brackish waters, have prominent growth rings, thickened shell margins, and a light neck with small dark markings. Once a popular gourmet food, they are still esteemed by epicures and produced on "farms."

DIAMONDBACK TERRAPIN (*Malaclemys terrapin*). Has broad, smooth crushing surfaces inside margins of jaws; its diet is almost exclusively snails, crabs, and other animals with hard outer skeletons. Also eats fish, worms, and other animals as it scavenges.

Seven subspecies: (1) *M. t. terrapin*—knobs on central laminae not expanded at tip, carapace wider toward rear; (2) *M. t. centrata*—knobs as in (1) but carapace equally wide at front and rear; (3) *M. t. tequesta*—knobs and carapace as in (2) but carapace is dark and center of large laminae is scarcely lighter; (4) *M. t. rhizophorarum*—knobs expanded at tip, laminae without light center, streaks on neck; (5) *M. t. macrospilota*—knobs as in (4), laminae with yellow-orange center; (6) *M. t. pileata*—knobs and laminae as in (4), body-skin dark; (7) *M. t. littoralis*—knobs and laminae as in (4), head and plastron white.

TORTOISES—family Testudinidae

Tortoises are strictly terrestrial, mostly dome-shelled turtles with stumplike or elephantine hind feet and scraperlike forefeet. All are fully shelled, with laminae over the shell. The tail is short, the bridge broad. Some of the well-developed plastral laminae, always 12 in number (6 pairs), are in contact with the marginal laminae of the carapace at the bridge. This group of 10 genera and 39 species is limited to warm regions. The three species in the United States have the northernmost distribution of tortoises.

GOPHER TORTOISES—genus *Gopherus*

The geographic ranges of the three species of gopher tortoises occurring in the United States are widely separated—identifiable by origin alone if it is known. One other species occurs in northwestern Mexico. The plastron projects forward under the head and neck. All gopher tortoises are largely vegetarian, eating flowers, fruits, leaves, and the stems of low-growing plants. In captivity, they can be fed any fruit or leaves that are acceptable also to humans. In addition, they will eat grass, dog food, raw meat, and hamburger. Gopher tortoises can go for long periods without drinking, instead using water in their food and that produced metabolically.

DIAMONDBACK TERRAPIN
female: 9 in. (229 mm)
male: 4½ in. (114 mm)

two male Desert Tortoises meet

Gopher Tortoise in burrow

DESERT TORTOISE (*Gopherus agassizi*). Has single axillary lamina at front edge of bridge. Front and hind feet are about equal in width. As in other species of tortoises, shell is dark brown, and in young, each lamina on carapace has a light center. Growth rings are evident on most of the laminae. Small skin gland is located on lower surface of each side of head; they produce a scent during courtship and also in combat. Like other tortoises, it is limited to sandy soils where it digs extensive burrows for protection and also for nesting. Eggs, which have shells that harden soon after they are laid, are put in nest 1–10 in. (25–250 mm) below surface. Eggs are somewhat elongate, 1.5 × 1.8 in. (38 × 46 mm), and 2 or 3 clutches of 2–14 eggs (usually 5–6) are laid every season.

The Desert Tortoise has separate feeding and hibernating areas. Though there is little evidence of territoriality, combats in which the front projection of the plastron is used as a ram are common when the tortoises encounter each other. Adults are completely docile when handled, but hatchlings are pugnacious. A protected species, the Desert Tortoise was formerly very popular as a pet.

GOPHER TORTOISE (*Gopherus polyphemus*). Has relatively small hind feet, and distance between bases of 1st and 4th claws of hind foot is about equal to distance between bases of 1st and 3rd claws of forefoot. There is usually a single axillary lamina. Like Desert Tortoise, the Gopher Tortoise digs extensive burrows. Its eggs are spherical, 1.5–1.8 in. (38–46 mm) in diameter. A protected species, severely threatened by its small range in an area densely populated by humans. A "key" species ecologically, with many others dependent upon it.

TEXAS TORTOISE (*Gopherus berlandieri*). Smaller and broader than other U.S. tortoises. Hind feet are relatively small, as in Gopher Tortoise, and there are usually 2 axillary laminae. Nest is very shallow—about 3 in. (76 mm) maximum—and eggs (only 1–3 per nest) are distinctly elongate, about 1.2 × 1.8 in. (30 × 46 mm). Rarely digs extensive burrows, usually seeking protection in shallow depressions called pallets. Uses its shell to scoop these out in the sand.

DESERT TORTOISE
14½ in. (368 mm)

GOPHER TORTOISE
14½ in. (368 mm)

TEXAS TORTOISE
8⅝ in. (219 mm)

SCALED REPTILES—order Squamata

This order is comprised of lizards, suborder Lacertilia (Sauria); snakes, suborder Serpentes (Ophidia); and amphisbaenids, suborder Amphisbaenia. Some authorities give all three groups order rank. All U.S. lizards possess either limbs or movable eyelids—or both. Snakes lack limbs and movable eyelids. Amphisbaenids have a short blunt tail, not tapering; an elongate body with earthwormlike rings; no ear openings; and eyes completely covered by scaly skin and usually not visible.

LIZARDS—suborder Lacertilia

Eight families of lizards of a total of about 20 worldwide are native to the United States. The family sequence in this book conforms with a phylogeny based primarily on brain structure.

KEY TO LIZARD FAMILIES

1. (*Not shown*) External limbs present
   . . . . *see* 3
   No external limbs . . . . *see* 2
2. Ear openings present; skin fold on each side of body . . . .
   . . . . **anguids,** p. 86
   Neither . . . . **legless lizards,** p. 90
3. All scales smooth, flat, rounded, subequal in size. . **skinks,** p. 72
   Not as above . . . . *see* 4
4. Movable eyelids present . . . *see* 6
   No movable eyelids . . . . . . *see* 5
5. Large plates on head; large quadrangular scales on belly . . . .
   . . . . **night lizards,** p. 84
   Tiny granules on head; small rounded scales on belly . . . . . .
   . . . . **geckos,** p. 66

6. (*not shown*) Dorsal and belly scales abruptly larger than lateral granules in longitudinal skin fold on sides of body . . **anguids,** p. 86
   Not as above . . . . *see* 7
7. Belly scales large, quadrangular, in 6–8 longitudinal series; all other scales granular . . . . *see* 8
   Not as above . . . . *see* 9
8. 1 large preanal . . . **lacertids,** p.82
   Several . . . . **teiids,** p. 92
9. Belly scales quadrangular . . . . . .
   . . . . **venomous lizards,** p. 92
   Rounded . . . . *see* 10
10. Scales under digits smooth; no parietal eye . . . . . **geckos,** p. 66
    Scales keeled; parietal eye present
    . . . . **iguanids,** p. 100

Geckos (Gekkonidae) are nocturnal lizards found worldwide in tropics and subtropics. Skinks (Scincidae) are diurnal, occurring worldwide in temperate and tropical regions. Night lizards (Xantusiidae) range from southwestern U.S. to Central America. Anguids (Anguidae) are diurnal and mostly American, absent or sparse elsewhere. Legless lizards (Anniellidae) are diurnal, restricted to Baja and southern California. Venomous lizards (Helodermatidae) are diurnal, restricted to Mexico and the southwestern U.S. Teiids (Teiidae), diurnal, are strictly American, temperate to tropical. Iguanids (Iguanidae), the largest family, are diurnal and almost exclusively American. In addition to native families, introductions of Lacertids (Lacertidae) from Europe have become established in North America.

KEY TO LIZARD FAMILIES

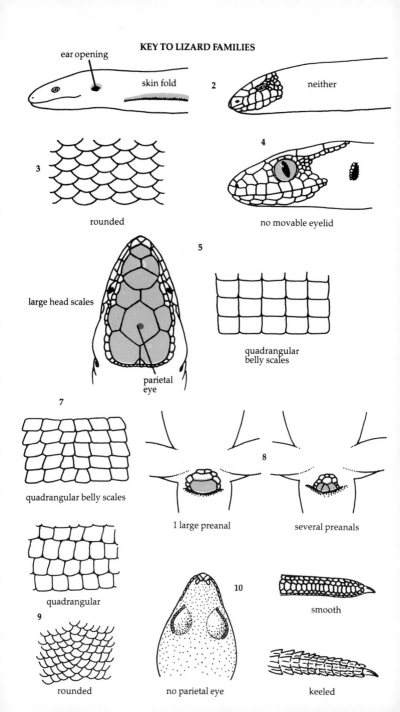

ear opening

skin fold

2

neither

3

rounded

4

no movable eyelid

5

large head scales

parietal eye

quadrangular belly scales

7

quadrangular belly scales

8

1 large preanal

several preanals

quadrangular

9

rounded

10

no parietal eye

smooth

keeled

GECKOS—family Gekkonidae

Geckos (85 genera, 650 species) occur worldwide in the tropics and subtropics. Except for three viviparous genera in New Zealand, all lay hard-shelled eggs. They are mostly nocturnal or crepuscular, and insectivorous. Three subfamilies in the United States.

GROUND GECKOS—subfamily Eublepharinae

Six genera and 16–18 species in Asia, Africa, and the Americas; in North America, 1 species of *Anarbylus* and 5 species of *Coleonyx*. Movable eyelids and padless feet are primitive characteristics. Preanal pores are well developed in males, rudimentary but visible in young and females. Pupils are vertical. All ground geckos are terrestrial, vocal, and lay two eggs only.

KEY TO BANDED GECKOS

1. (*not shown*) Back with scattered enlarged tubercles . . . . *see* **2**
   Uniform granular scales . . . *see* **3**
2. (*not shown*) Tail black and white banded, found in southern California . . . .
   . . . . **Baja Banded,** p. 68
   Bands not black and white, found in Texas . . . **Reticulated,** below
3. Males with preanal pores in 2 series separated by 1 or more median scales; postanal spur flat . .
   . . . . **Texas Banded,** below
   Preanal pores in continuous series, undivided; spur pointed . . . . .
   . . . . . **Western Banded,** below

TEXAS BANDED GECKO (*Coleonyx brevis*). Has 3–6 preanal pores. Young have bright crossbands wider than spaces between, and light neck crescent. Crossbands fade to gray-brown with small dark spots in adults.

WESTERN BANDED GECKO (*Coleonyx variegatus*). Has 4–10 preanal pores. Crossbands not so wide as spaces between in young; in larger animals, dark spots become paler in and between; in largest, borders may break up into dark spots. Four subspecies in U.S.: (1) *C. v. variegatus*—crossbands little wider than spaces between, head spotted, preanal pores seldom 8 or more; (2) *C. v. abbotti*—like (1) except head unspotted; (3) *C. v. bogerti*—like (1) but usually 8 or more preanal pores in males; (4) *C. v. utahensis*—like (1) but crossbands wider than spaces.

RETICULATED GECKO (*Coleonyx reticulatus*). Has 20–24 preanal pores. Scattered enlarged tubercles. Young have 4 dark crossbands, not so wide as spaces between; crossbands become lighter in adults, with scattered black spots except on midline.

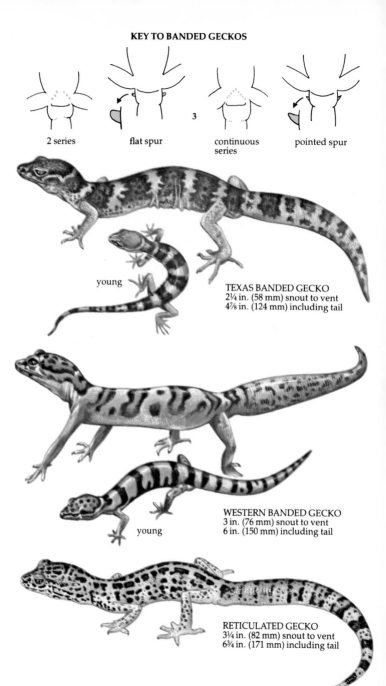

KEY TO BANDED GECKOS

2 series

flat spur

3

continuous series

pointed spur

TEXAS BANDED GECKO
2¼ in. (58 mm) snout to vent
4⅞ in. (124 mm) including tail

young

WESTERN BANDED GECKO
3 in. (76 mm) snout to vent
6 in. (150 mm) including tail

young

RETICULATED GECKO
3¼ in. (82 mm) snout to vent
6¾ in. (171 mm) including tail

BAJA BANDED GECKO (*Anarbylus switaki*). A large gecko, 3¼ in. (83 mm) snout to vent, with enlarged tubercular scales on back; tail black and white banded. Bases of claws enclosed by sheath of scales; digits have granules rather than broad scales below. Rare, in southern California and Baja California. [Not illustrated; no range map.]

TRUE GECKOS—subfamily Gekkoninae

The 60 genera and 400 species of true geckos occur in tropical and subtropical regions of the world and have been widely introduced along shipping routes. All species in the U.S. are introduced except the Leaf-toed Gecko. This subfamily is characterized by fused and transparent eyelids, large digital pads, and vertical eye pupils with scalloped margins. True geckos are vocal, deposit 2 eggs per clutch, and usually have postcloacal bones. Excellent climbers, they are often abundant in human habitations crawling on walls and ceilings, the numerous slender filaments on the underside of their toe pads catching on irregularities. They are insectivorous, and nocturnal or crepuscular. They clean their eyes by licking them with their flat tongue. In most species, preanal or femoral pores, or both, are well developed in mature males; rudimentary in juveniles and females.

KEY TO TRUE GECKOS

1. 2 pads at tips of digits . . . . **Leaf-toed**, p. 70
   Digits expanded but no pads at tips . . . . *see* **2**
2. No enlarged skin tubercles . . . .**Stump-toed**, below

   At least 1 row . . . . *see* **3**
3. Tubercles large, numerous . . . . **Mediterranean**, p. 70
   Tubercles small, 1 row per side . . . .**Fox**, below

STUMP-TOED GECKO (*Gehyra mutilata*). Tiny, granular dorsal scales. Median row on underside of tail enlarged. Both preanal and femoral pores. Tail flattened, and edges saw-toothed. Gray to pinkish, sometimes flecked, often with dark-edged spots and a dark streak behind the eye; whitish below. Occurs in San Diego, Calif. [No map.]

FOX GECKO (*Hemidactylus garnoti*). Granular dorsal scales except 1 row of enlarged rounded tubercles on each side. Tail is strongly depressed, with edges saw-toothed. Rear pair of chin shields is separated from labials. Parthenogenetic (no males). Rudimentary femoral pores, no preanals. Occurs in Miami, Florida. *H. frenatus*, introduced widely into western Mexico, is expected in adjacent U.S. Weakly depressed tail without saw-toothed edges; all chin shields contact labials.

geckos clean their transparent
eye scales by licking them

note large digital pads

STUMP-TOED GECKO
2⅜ in. (62 mm) snout to vent
4¾ in. (121 mm) including tail

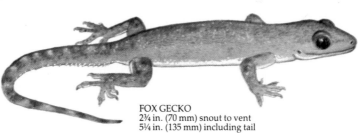

FOX GECKO
2¾ in. (70 mm) snout to vent
5¼ in. (135 mm) including tail

MEDITERRANEAN GECKO (*Hemidactylus turcicus*). Mostly granular scales; numerous enlarged tubercles in roughly 14–16 rows, some extending onto tail. Preanal pores only, no femorals. Pale pinkish to gray-brown with scattered irregular dark marks; whitish below. Nocturnal. One subspecies in N.A.: *H. t. turcicus*.

The similar *H. mabouia* has been introduced widely to West Indies and eastern Mexico; spread to U.S. expected.

LEAF-TOED GECKO (*Phyllodactylus xanti*). A single pair of enlarged pads at tip of each digit (unique among N.A. geckos). Dorsal scales are mostly granular, with numerous enlarged keeled tubercles among them. Tail is not flattened. No preanal or femoral pores. Gray to yellowish white, translucent in life; 7-8 crossbands on back; sides, limbs, and head mottled; dark streak on sides. One subspecies in U.S.: *P. x. nocticolus*.

DWARF GECKOS—subfamily Sphaerodactylinae

The 5 genera and about 95 species of dwarf geckos are limited to the American tropics, including the southern tip of Florida where they were apparently introduced. *Gonatodes* is diurnal, with round pupils; *Sphaerodactylus* is crepuscular, with elliptical pupils. Dwarf geckos have no preanal or femoral pores, voice, cloacal bones. They lay 1 egg per clutch. Eyelids are fused and transparent, forming "spectacles" (as in true geckos). Digital pads, if present, are small. Dwarf geckos are good climbers but cannot crawl upside down. Insectivorous.

KEY TO DWARF GECKOS

1. Digital tips expanded . . . . . *see 2*
   Not expanded . . . .
   . . . . . . **Yellow-headed,** below
2. Dorsal scales granular . . . .
   . . . . **Ashy,** p. 72

Overlapping, keeled . . . . . . *see 3*
3. Numerous white dots on neck
   . . . . **Ocellated,** below
   2 white dots or none . . **Reef,** p. 72

YELLOW-HEADED GECKO (*Gonatodes albogularis*). Granular dorsal scales. Young and females mottled gray-brown, with narrow light collar. Males uniformly gray or black, except for yellow head and neck. Usually light below, but dark in old males. Tip of tail whitish (unless regenerated). One subspecies in U.S.: *G.a. fuscus*.

OCELLATED GECKO (*Sphaerodactylus argus*). Smaller dorsal scales than Reef, 57–73 at midbody. Dark streaks prominent in young, lost in old adults. Tail reddish; white dots on nape. One U.S. subspecies: *S. a. argus*.

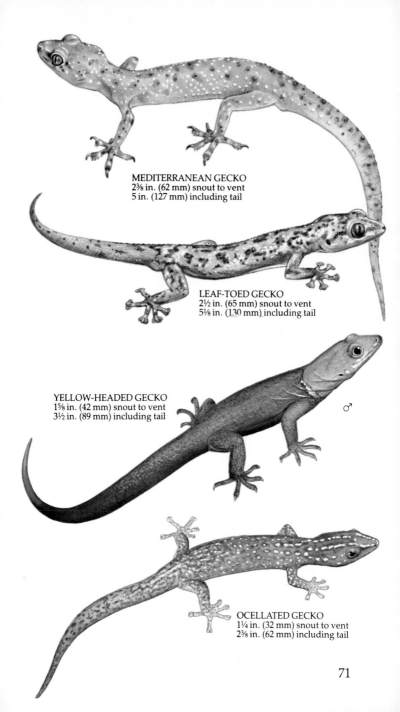

MEDITERRANEAN GECKO
2⅜ in. (62 mm) snout to vent
5 in. (127 mm) including tail

LEAF-TOED GECKO
2½ in. (65 mm) snout to vent
5⅛ in. (130 mm) including tail

YELLOW-HEADED GECKO
1⅝ in. (42 mm) snout to vent
3½ in. (89 mm) including tail

♂

OCELLATED GECKO
1¼ in. (32 mm) snout to vent
2⅜ in. (62 mm) including tail

71

ASHY GECKO (*Sphaerodactylus cinereus*). Young are gray to reddish-brown, with 5 dark, narrow crossbands between fore and hind limbs; ends of limbs and tail are reddish. Adults are uniformly brownish with numerous yellow or white dots, which are largest and tend to form rows on head and tail. U.S. subspecies: *S. c. elegans*.

REEF GECKO (*Sphaerodactylus notatus*). Has 41–48 scales around midbody. Brownish to red-brown. Young and females have numerous dark flecks, which approach rows on head and neck. Males are uniformly dotted. Larger dorsal scales than in Ocellated Gecko. One subspecies in U.S.: *S. n. notatus*.

SKINKS—family Scincidae

The third largest family of lizards (about 50 genera, 550 species), skinks are cosmopolitan in distribution except at high latitudes. They are the dominant lizards of the Pacific, where they are readily rafted between islands. Diurnal and primarily terrestrial, a few are semiarboreal, semi-fossorial, or semiaquatic. Most are insectivorous; some prey on small vertebrates, and at least a few are largely herbivorous. Most are oviparous, with females of some species remaining with their eggs until hatching, but many are viviparous, with newborn one-half the length of the adult. In the most primitive skinks, the eyelid is covered by numerous scales; in more advanced species, a large translucent scale serves as a "window" when the lid is closed; in the most advanced species, the eyelid is permanently fused and the window scale covers the entire eye. Burrowing habits of the family have probably exerted strong selective pressure favoring protection of the eyes and, in several species groups, have also led to loss of ear openings and reduction or loss of limbs. In the United States only Sand Skinks (*Neoseps*) show both these trends; all others have normal limbs and visible ear openings. There are no femoral pores, and the sexes are difficult to distinguish. Most skinks have readily breakable tails which, at least in juveniles, are brightly colored; the bright, actively wriggling broken tail is thought to attract predators, allowing the tailless but otherwise unharmed lizard to escape.

KEY TO SKINKS

1. Legs short, no longer than head; no more than 2 digits; lower eyelids with translucent window . . . . **Sand,** p. 80
   Legs normal, 5 digits . . . . . *see* **2**

2. Pair of scales bordering rostral; lower eyelids scaly . . . .**striped skinks,** p. 74
   1 scale bordering rostral; lower eyelids with translucent window . . . . **Ground,** p. 80

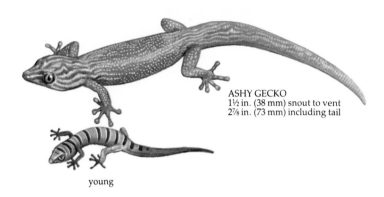

ASHY GECKO
1½ in. (38 mm) snout to vent
2⅞ in. (73 mm) including tail

young

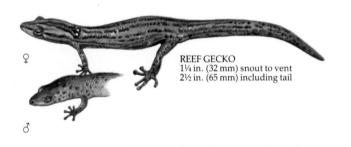

♀

REEF GECKO
1¼ in. (32 mm) snout to vent
2½ in. (65 mm) including tail

♂

KEY TO SKINKS

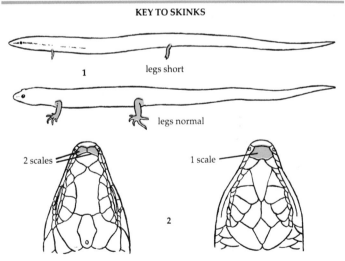

1

legs short

legs normal

2 scales

1 scale

2

STRIPED SKINKS—genus *Eumeces*

This is the most primitive genus of the family, widely distributed in both hemispheres but largely northern. It includes about 35 species, 12 in U.S. These skinks have opaque, windowless eyelids; strong limbs; and smooth, rounded scales, the median pair behind the rostral unique to this genus. All in U.S. are oviparous. Most U.S. species are terrestrial; all burrow.

KEY TO STRIPED SKINKS

1. Some rear lateral scale rows oblique . . . . **Great Plains,** p. 80
   All scales in parallel rows . . *see* **2**
2. 1 scale between last labial and parietal . . . . **Mole,** p. 80
   3 scales . . . . *see* **3**
3. 1 postmental . . . . *see* **4**
   2 postmentals . . . . *see* **7**
4. Median row of subcaudals scarcely or not widened where tail unregenerated . **Four-lined,** p. 76
   Distinctly widened . . . . *see* **5**
5. (*not shown*) No dorsolateral lines, or on 3rd scale row only . . . . **Many-lined,** p. 78
   Lines present, not on 3rd row only . . . . *see* **6**
6. (*not shown*) Light line on edges of 4th or 4th and 5th scale rows . . . . **Prairie,** p. 76
   On 3rd and 4th rows . . **Coal,** p. 76
7. Supranasal contacting 2 scales behind nasal (postnasal present) . . . . *see* **10**
   Contacting 1 scale behind nasal (no postnasal) . . . . *see* **8**
8. Parietals contacting behind interparietals . . . . . **Mountain,** p. 76
   Not contacting . . . . *see* **9**
9. (*not shown*) Light line following 4th or 4th and 5th scale rows . . . . **Prairie,** p. 76

   No line, or on 3rd scale row only . . . . **Many-lined,** p. 78
10. (*not shown*) Unicolor above, no lines . . . . *see* **11**
    Light lines visible . . . . *see* **12**
11. E of 98th meridian . . . . *see* **16**
    W of 109th meridian . . . . **Gilbert's,** p. 78
12. E of 98th meridian . . . . *see* **16**
    W of 109th meridian . . . . . *see* **13**
13. (*not shown*) Line on 3rd scale row only . . . . . . **Many-lined,** p. 78
    On 2nd and 3rd rows . . . . . *see* **14**
14. (*not shown*) Tail red to pink . . . .**Gilbert's,** p. 78
    Tail bluish . . . . *see* **15**
15. (*not shown*) Scales in light lines edged with brown or gray . . . .**Gilbert's,** p. 78
    Not as above . . . . **Western,** p. 78
16. (*not shown*) Median scale row under tail scarcely wider than adjacent rows (where unregenerated) . . . . **Southeastern Five-lined,** below
    Row much wider . . . . *see* **17**
17. No postlabials, or 1 or 2 small . . . . **Broadhead,** p. 76
    2 postlabials, relatively large . . . .**Five-lined,** p. 76

Five-lined Group: These lizards start life five-lined and blue-tailed. Tail later turns gray or brown, matching body. Light lines largely or entirely disappear in males, fade in females; median light line is forked on head. All have 1 postnasal and 2 postmentals.

SOUTHEASTERN FIVE-LINED SKINK (*Eumeces inexpectatus*). Has 4 or 5 upper labials preceding subocular; median scale row under tail scarcely or not enlarged where tail not regenerated. Light lines narrow; dorsolaterals on 5th or 4th and 5th scale rows from midline.

KEY TO STRIPED SKINKS

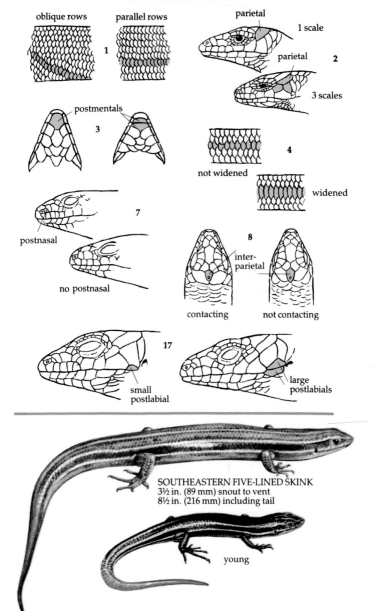

oblique rows parallel rows **1**

parietal

1 scale

parietal

3 scales

2

postmentals

3

not widened

widened

4

postnasal

7

no postnasal

8

inter-parietal

contacting not contacting

17

small postlabial

large postlabials

SOUTHEASTERN FIVE-LINED SKINK
3½ in. (89 mm) snout to vent
8½ in. (216 mm) including tail

young

BROADHEAD SKINK (*Eumeces laticeps*). Usually has 5 (often 4) upper labials preceding subocular; median scale row under tail much wider than adjacent rows. Light lines usually on 4th row, at times 3rd and 4th rows, from midline. Extensively arboreal.

FIVE-LINED SKINK (*Eumeces fasciatus*). Usually has 4 (occasionally 5) upper labials; tail as in Broadhead. Light lines on edges of rows 3 and 4.

Short-lined Group: Young are five-lined with median line forked on head; median line is always short and disappears or dims in adults. Other lines also become dim, short, or broken. Postnasal regularly absent; 4 upper labials precede the subocular.

MOUNTAIN SKINK (*Eumeces callicephalus*). Has 2 postmentals; median subcaudals are slightly widened. Broad dark line between light lines on sides; upper light line on 4th row only, not beyond midbody in adults. Tail is bright blue in young, not as bright in adults.

FOUR-LINED SKINK (*Eumeces tetragrammus*). Has 1 postmental (rarely 2); median subcaudals scarcely or not enlarged. Light line on edges of rows 3 and 4. In young, sides darker, light lines orange, tail azure. Two subspecies: (1) *E. t. tetragrammus*—dorsolateral line to hind legs; (2) *E. t. brevilineatus*—to midbody.

Eastern Four-lined Group: Three species are known, one restricted to Mexico. No median or forked light line on head, rarely a postnasal. Median subcaudals are enlarged. Male's head is reddish in breeding season. Tail blue in young. Light stripes extend onto tail in all.

COAL SKINK (*Eumeces anthracinus*). Has 1 postmental (rarely 2), light stripes on rows 3 and 4, sometimes a median light line, dark lateral stripe 2.5–4 scales wide. Two subspecies: (1) *E. a. anthracinus*—25 or fewer scale rows, young patterned like adults; (2) *E. a. pluvialis*—26 or more scale rows, young black, sometimes with trace of light lines.

PRAIRIE SKINK (*Eumeces septentrionalis*). Has 2 postmentals (rarely 1), light stripes on row 4 or rows 4 and 5, dark lateral stripe 2 scales wide. Two subspecies: (1) *E. s. septentrionalis*—several dark lines bordering light; (2) *E. s. obtusirostris*—reduced or no dark lines.

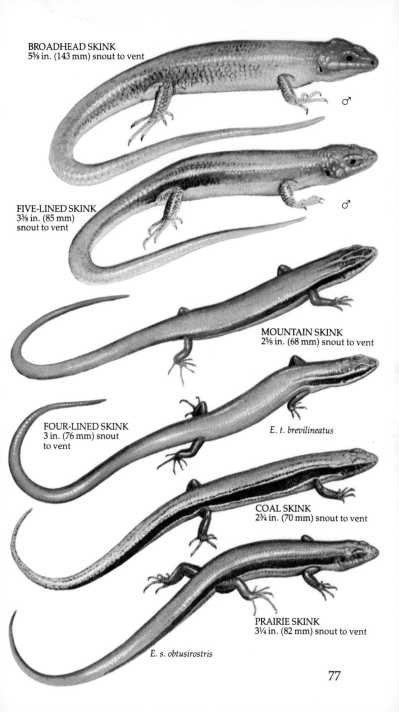

BROADHEAD SKINK
5⅝ in. (143 mm) snout to vent

♂

FIVE-LINED SKINK
3⅜ in. (85 mm)
snout to vent

♂

MOUNTAIN SKINK
2⅝ in. (68 mm) snout to vent

FOUR-LINED SKINK
3 in. (76 mm) snout
to vent

E. t. brevilineatus

COAL SKINK
2¾ in. (70 mm) snout to vent

PRAIRIE SKINK
3¼ in. (82 mm) snout to vent

E. s. obtusirostris

77

Western Four-lined Group: These skinks have broad dorsolateral and lateral light lines on the second and third scale rows from the midline and extending onto the tail, at least in the young. Adults of one species lose all evidence of lines. There is no median line, but always a post-nasal and two postmentals. Median subcaudals are greatly widened.

WESTERN SKINK (*Eumeces skiltonianus*). Usually has 7 supralabials and 4 nuchals. Tail is bright blue in young, dull blue or gray in adults. In breeding season, orange appears on sides of head, on chin, on underside, and on tip of tail. Three subspecies: (1) *E. s. skiltonianus*—light stripe no more than half width of 2nd scale row, at mid-body less than half the width of middorsal stripe; (2) *E. s. utahensis*—light stripe more than half the width of 2nd scale row, at midbody half or more the width of middorsal stripe: (3) *E. s. interparietalis*—parietals in contact behind interparietals, median and lateral dark stripes to or beyond middle of tail (not in others).

GILBERT'S SKINK (*Eumeces gilberti*). Usually has 8 supralabials, 2–4 nuchals. Like the Western Skink, but in young, dark stripe ends at base of tail; most adults are unstriped, with tail brick-red or orange. Five subspecies: (1) *E. g. gilberti*—young blue-tailed, 2 pairs of nuchals; (2) *E. g. placerensis*—like (1) but with 1 pair of nuchals; (3) *E. g. cancellosus*—young pink-tailed, tinged with blue above, more mature with dark bars or latticework; (4) *E. g. rubricaudatus*—young pink-tailed, no blue, markings lost earlier than in (3); (5) *E. g. arizonensis*—young pinkish only under tail, adults retain evidence of stripes.

Miscellaneous Groups: These three distinctive species of skinks are not closely related to the other United States *Eumeces*.

MANY-LINED SKINK (*Eumeces multivirgatus*). Rarely has 1 postmental; usually has a postnasal (often none). Median subcaudals are wide. Pattern variable, with some adults uniformly brownish. Light lines are narrow, on 3rd scale row from midline; dim or broken on 6th scale row. Tail is blue or lavender in young, body dark except for primary light lines. Two subspecies: (1) *E. m. multivirgatus*—no median light line in young, broad and dark-edged in adults; other dark and light lines on back and sides; (2) *E. m. gaigeae*—median light line in young, adults often unlined.

WESTERN SKINK
3¼ in. (82 mm) snout to vent
9½ in. (241 mm) including tail

GILBERT'S SKINK
4½ in. (114 mm) snout to vent
13 in. (330 mm) including tail

MANY-LINED SKINK
2⅞ in. (73 mm) snout to vent
7⅝ in. (194 mm) including tail

GREAT PLAINS SKINK (*Eumeces obsoletus*). Diagonal rows of scales on lower rear sides, not parallel with dorsals; rarely 1 postmental; with or without postnasal. Median subcaudals are wide. Scales are light gray to tan edged with black, often forming jagged lines. Young are jet-black with orange and white dots on the head; tail is bright blue.

MOLE SKINK (*Eumeces egregius*). This is the most highly specialized species of the genus in our area. One scale between rear upper labial and parietal; no postnasal; 2 postmentals; 3 supraoculars (4 in other *Eumeces*); 18–24 scale rows at midbody; 2 middorsal scale rows; median subcaudal scale row widened. Limbs slender and short; ear opening small, and partially closed. Light stripe on 2nd row or 2nd and 3rd rows sometimes reaches tail. May or may not have lateral light line, never beyond groin.

Five subspecies: (1) *E. e. egregius*—tail red, dorsolateral light lines not widening, lateral line conspicuous; (2) *E. e. similis*—tail red, 6 upper labials (others have 7); (3) *E. e. insularis*—tail reddish, young black, light lines scarcely discernible and not widened; (4) *E. e. onocrepis*—tail yellow to red, brown, or violet (never blue), dorsolateral lines short, widened or diverged or both; (5) *E. e. lividus*—tail blue, dorsolateral lines as in (4).

GROUND SKINKS—genus *Scincella*

GROUND SKINK (*Scincella lateralis*). Has 1 median frontonasal contacting rostral, no supranasals; a pair of scales following frontal; window in lower eyelid. Scales are smooth, cycloid, and equal in size; 26-32 rows at midbody. Adpressed limbs are separated by 10-20 scales. Oviparous.

SAND SKINKS—genus *Neoseps*

SAND SKINK (*Neoseps reynoldsi*). Hind legs are about as long as head, with 2 digits; forelegs half as long, with 1 digit. Eyes are small; lower eyelid has large window scale. No ear openings. Head is flat below, rounded above; snout rounded, sharp-edged; lower jaw flush with upper. All scales are smooth and cycloid, in 16 rows; 2 middorsal rows widest. There is a sharply defined keel on each side of belly; body is flat or concave below. Oviparous.

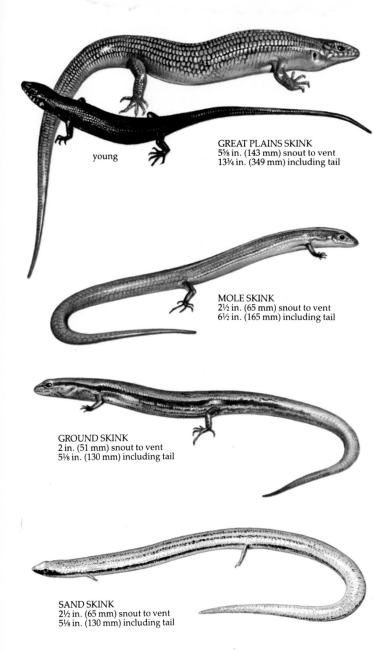

young

GREAT PLAINS SKINK
5⅝ in. (143 mm) snout to vent
13¾ in. (349 mm) including tail

MOLE SKINK
2½ in. (65 mm) snout to vent
6½ in. (165 mm) including tail

GROUND SKINK
2 in. (51 mm) snout to vent
5⅛ in. (130 mm) including tail

SAND SKINK
2½ in. (65 mm) snout to vent
5⅛ in. (130 mm) including tail

LACERTIDS—family Lacertidae

Lacertids are not native to the Western Hemisphere, but three separate introductions of the genera *Lacerta* and *Podarcis* from southern Europe are known to have taken place. All have been in densely populated areas, hence limiting expansion and increasing the likelihood of extinction. Populations of the Italian Wall Lacerta in West Hempstead, Long Island (1966), and the Green Lacerta in Topeka, Kansas (1968), persist and reproduce.

Lacertids form a family of more than 20 genera and 175 species native to Europe, Africa, Asia, and the East Indies. The genus *Lacerta* (about 35 species) occurs in Europe, Asia, and northern Africa. *Podarcis* (about 12 species) occurs in the Mediterranean region. *Lacerta* and *Podarcis* closely resemble the American genera *Cnemidophorus* and *Ameiva*, but the families Lacertidae and Teiidae have widely different evolutionary histories.

In the Green Lacerta, median rows of ventrals overlap with oblique lateral edges of the scales; in the Italian Wall Lacerta they meet in a straight midventral line.

ITALIAN WALL LACERTA (*Podarcis sicula*). Resembles a whiptail (*Cnemidophorus*)—body and tail elongate, head pointed, limbs short but strong. Dorsals granular, uniform in size, feebly keeled; head scales large and flat; ventrals rectangular, large, flat, and smooth, in 6 rows; granular gular fold bordered anteriorly by large scales; femoral pores, 16–26 per side, separated by 1 median scale. This terrestrial insect-eater is oviparous. Two subspecies in U.S. apparently freely interbreed: (1) *P. s. sicula*—green on midback only, mostly yellowish gray above, 3 dorsal rows of dark spots; (2) *P. s. campestris*—mostly green above, rows of dark spots often fused as continuous lines. [No range map.]

GREEN LACERTA (*Lacerta viridis*). Relatively large scales between eye and ear, a row of tiny scales between 2 supraoculars and superciliaries. Dorsal scales are keeled and granular; those in 1–2 vertebral rows are smaller, and there are 42–56 in a transverse row. Ventrals are in 6–8 rows. Adults greenish above, whitish below except throat and sides of head blue, especially in males during breeding season. Males black-speckled above, females blotched or with 2 or 4 light stripes. Young brownish with 2–4 light stripes. Eats insects, smaller lizards, and rodents. Subspecies in U.S. uncertain. [No range map.]

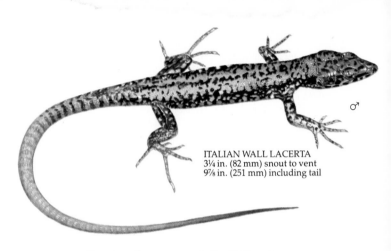

ITALIAN WALL LACERTA
3¼ in. (82 mm) snout to vent
9⅞ in. (251 mm) including tail

♂

Green Lacerta   Italian Wall Lacerta

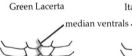

median ventrals

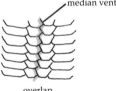

overlap   in straight line

♂

young

GREEN LACERTA
5⅛ in. (130 mm) snout to vent
15¾ in. (400 mm) including tail

NIGHT LIZARDS—family Xantusiidae

Five genera and 16 species occur from the southwestern United States to Panama and Cuba. Nocturnal and secretive, these lizards have fixed transparent eyelids that look like the eye itself, vertical pupils, and conspicuous ear openings. Other characteristics are large plates on the head, granular dorsal scales, large rectangular ventral scales, a granular gular fold, and no osteoderms. The limbs and tail are normal, with femoral pores present. The skin is soft and fragile, the tail easily broken. Night lizards are viviparous. The three United States species are distinguished by the number of longitudinal rows of ventrals: the Island Night Lizard has 16; the Granite, 14; and the Desert, 12. There are 5–12 femoral pores, and the tail is fragile.

ISLAND NIGHT LIZARD (*Klauberina riversiana*). Has 2 rows of small scales over eye; 6–8 preanal scales, small and irregular; skin loose, with vertical folds on sides. Color is gray to cinnamon, with dark brown speckling or reticulation, coarse or fine. Often dark-edged light lines are present on the back with median dark area. White or dusky below, with small black flecks at sides of the throat, on the belly, and on the tail. Restricted to San Nicolas, Santa Barbara, and San Clemente islands off southern California; insular differences may be of subspecific rank.

GRANITE NIGHT LIZARD (*Xantusia henshawi*). This lizard has 1 row of small scales over eye; 1 large and 1 small pair of preanals; skin loose, with vertical folds on sides. Yellowish to brown above, with numerous large dark spots. The head is uniformly dark or with light-edged scales. Limbs and tail are spotted and mottled. Cream-colored and unmarked below. One subspecies in U.S.: *X. h. henshawi*.

DESERT NIGHT LIZARD (*Xantusia vigilis*). This species is velvet-skinned. It has 1 row of small scales over eye, and 2 pairs of preanals of equal size. Four subspecies in U.S.: (1) *X. v. vigilis*—33–40 longitudinal rows of dorsals at midbody, 18–21 lamellae under 4th toe; (2) *X. v. utahensis*—like (1) except 23–25 lamellae under 4th toe; (3) *X. v. arizonae*—42–50 rows of dorsals, 25–28 lamellae; (4) *X. v. sierrae*—40–44 rows of dorsals, 22–25 lamellae.

84

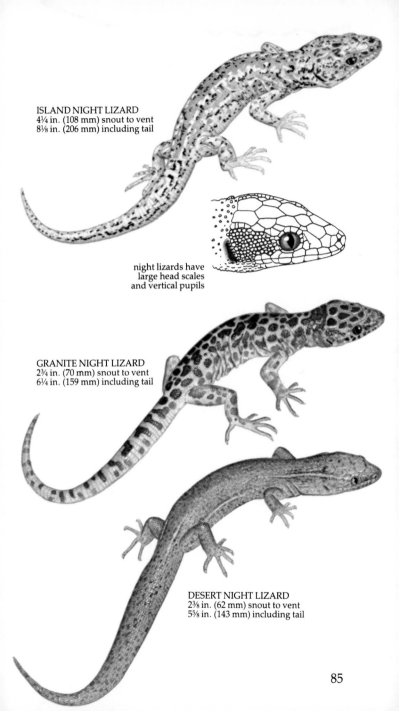

ISLAND NIGHT LIZARD
4¼ in. (108 mm) snout to vent
8⅛ in. (206 mm) including tail

night lizards have
large head scales
and vertical pupils

GRANITE NIGHT LIZARD
2¾ in. (70 mm) snout to vent
6¼ in. (159 mm) including tail

DESERT NIGHT LIZARD
2⅜ in. (62 mm) snout to vent
5⅝ in. (143 mm) including tail

85

ANGUIDS—family Anguidae

This family of 12 genera and 95 species occurs in temperate and tropical regions of the Americas, Europe, northern Africa, southern Asia, Taiwan, Sumatra, and Outer Hebrides. Diagnostic features are internal. A thick tongue and bony plates supporting most scales are present. Some anguids are smooth-scaled and skinklike; others are long-bodied and limbless, resembling snakes; still others are strong-limbed climbers. Of the three living subfamilies, all North American species are in Ophisaurinae (six genera), distinguished by a lateral granular fold separating platelike dorsal and ventral scales. This fold allows enlargement of the body for egg development and ingestion of large meals. All species in our area readily lose their tails and bite aggressively.

Of the three genera in North America, the glass lizards are limbless, the others are four-limbed. Texas alligator lizards have a median postrostral, and the nasals are separated from the rostral. Pacific alligator lizards lack a median postrostral, and the nasals contact the rostral.

PACIFIC ALLIGATOR LIZARDS—genus *Elgaria*

Six species comprise this genus, limited to the Pacific slopes of the United States, Canada, and northwestern Mexico; four occur north of Mexico. Both dorsals and ventrals are quadrangular, with ventrals always in 12 rows. Oviparous except Northern.

KEY TO PACIFIC ALLIGATOR LIZARDS

1. Dark crossbands occupying more than 2 rows of scales . . . . *see* **2**
   Fewer than 2 rows . . . . *see* **3**
2. Lips barred . . . .**King,** below
   Lips unicolor, light . . . .
   . . . . **Panamint,** below
3. Dark lines down middle of ventral scale rows; crossbands clearly evident; oviparous . . . .
   . . . . **Southern,** p. 88
   Dark lines down edges of ventral scale rows; crossbands indistinguishable; viviparous . . . .
   . . . . **Northern,** p. 88

KING ALLIGATOR LIZARD (*Elgaria kingi*). Has 14 (sometimes 16) dorsal scale rows, 50–62 scales in row from head to above anus; median 6–8 rows feebly keeled; 8–11 dark crossbands (which are bright in young), broader than the pale spaces between them and darker on rear edges. Lips are barred; eyes are orange or pink. One subspecies in U.S.: *E. k. nobilis.*

PANAMINT ALLIGATOR LIZARD (*Elgaria panamintina*). Much like King, but has 7–8 crossbands which are dark on both edges. Lips are light, not barred; eyes are pale yellow.

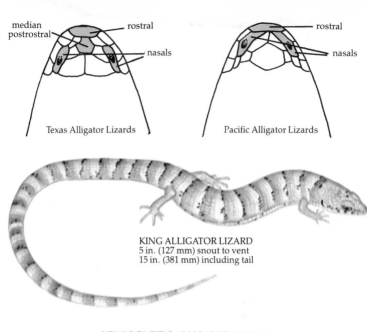

median postrostral · rostral · nasals

Texas Alligator Lizards

rostral · nasals

Pacific Alligator Lizards

KING ALLIGATOR LIZARD
5 in. (127 mm) snout to vent
15 in. (381 mm) including tail

KEY TO PACIFIC ALLIGATOR LIZARDS

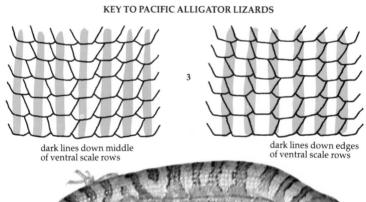

dark lines down middle
of ventral scale rows

3

dark lines down edges
of ventral scale rows

PANAMINT ALLIGATOR LIZARD
6 in. (150 mm) snout to vent
18 in. (456 mm) including tail

SOUTHERN ALLIGATOR LIZARD (*Elgaria multicarinata*). Has keeled dorsal scales in 14 rows (rarely 12 or 16), 40–52 scales from head to above anus. Brown, gray, yellowish, or reddish above. Crossbands, 9–13, from head to hind legs. Eyes pale yellow. Dim dark streak down middle of each ventral scale row. Young have broad, light brown band, sides crossbanded, vertebral row of small dots. Three subspecies: (1) *E. m. multicarinata*—head mottled (not in others), upper arm with 1–3 weakly keeled scale rows, temporals smooth; (2) *E. m. scincicauda*—no keels on arm, temporals smooth or upper ones faintly keeled; (3) *E. m. webbi*—at least 3 rows of upper arm scales keeled, at least upper temporals strongly keeled.

NORTHERN ALLIGATOR LIZARD (*Elgaria coerulea*). Like Southern but has shorter tail. Crossbands broken, indiscernible. Olive, greenish, or bluish. Eyes dark and dark streaks along edges of ventral scale rows. Four subspecies: (1) *E. c. coerulea*—scales strongly keeled (not in others), 14–16 rows; (2) *E. c. shastensis*—dorsals in 16 rows, temporals smooth; (3) *E. c. palmeri*—dorsals in 16 rows, temporals all keeled; (4) *E. c. principis*—dorsals in 14 rows, weakly keeled.

TEXAS ALLIGATOR LIZARDS—genus *Gerrhonotus*

One species occurs in the genus, widespread in Mexico. Head is more elongate and limbs weaker than in Pacific alligator lizards. Oviparous.

TEXAS ALLIGATOR LIZARD (*Gerrhonotus liocephalus*). Has 16 rows of rectangular dorsal scales with the median 8 rows distinctly keeled, next rows feebly keeled, others smooth. Ventrals smooth, in 12 rows. One subspecies in U.S.: *G. l. infernalis*.

GLASS LIZARDS—genus *Ophisaurus*

Eleven species comprise this genus, occurring in eastern North America (three north of Mexico), Europe, Asia, Africa, Sumatra, and Borneo. All are limbless (some Old World species have tiny hind limbs) and have large quadrangular scales, small ear openings, eyes with movable lids, and a long tail (to twice the head-and-body length). Oviparous.

KEY TO GLASS LIZARDS

1. Dark marks below lateral fold on neck . . . . **Slender**, p. 90
   No dark marks . . . . *see* **2**

2. Some upper labials reach orbit. . . . . **Island**, p. 90
   None . . . . **Eastern**, p. 90

SOUTHERN ALLIGATOR LIZARD
6½ in. (165 mm) snout to vent
20 in. (508 mm) including tail

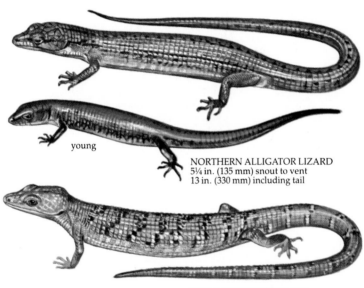

young

NORTHERN ALLIGATOR LIZARD
5¼ in. (135 mm) snout to vent
13 in. (330 mm) including tail

TEXAS ALLIGATOR LIZARD
8 in. (203 mm) snout to vent
20 in. (508 mm) including tail

tail breaks easily

contacting orbit

2

not contacting orbit

lateral fold

 KEY TO GLASS LIZARDS

ISLAND GLASS LIZARD (*Ophisaurus compressus*). Smallest U.S. glass lizard. Tail not especially brittle. Scales along lateral fold are 97 or fewer; frontonasal usually divided. Yellowish above with middorsal stripe, sometimes broken, and dark stripe along scale rows 3 and 4. Numerous white bars on neck. Pinkish buff or yellowish and unmarked below.

EASTERN GLASS LIZARD (*Ophisaurus ventralis*). Has an extremely brittle tail; 98 or more lateral fold scales; frontonasal usually undivided. Young pale brown, with broad, dark lateral stripe. Adults greenish, with dark streaks on all dorsal scale rows between lateral folds; indistinct or no middorsal stripe. Yellowish, unmarked below, with vertical white bars on neck.

SLENDER GLASS LIZARD (*Ophisaurus attenuatus*). Extremely brittle tail. No labials reach orbit; lateral fold scales number 98 or more; frontonasal usually undivided. Pale brown, with middorsal stripe sometimes broken, and several lateral dark stripes. Crossbands found in some adults; some vertical white streaks on neck. Two subspecies: (1) *O. a. attenuatus*—intact tail less than 2.4 times snout-to-vent length; (2) *O. a. longicaudus*—intact tail 2.4 times or more snout-to-vent length.

LEGLESS LIZARDS—family Anniellidae

This family consists of only one genus and two species (one in the United States). These lizards range from sea level to 6,000 feet (1,828 m) and are highly adapted for burrowing in sandy or loam soils and in leaf litter—wherever moisture is available just under the surface. Nocturnal, they come to the surface at dusk and night. Scales are smooth, round, and equal in size; eyes are small but with lids; there are no ear openings; the snout is shovel-shaped and the lower jaw counterset; the tail is short, its tip rounded and blunt. These lizards feed on insects and spiders. They bear 1–4 young.

CALIFORNIA LEGLESS LIZARD (*Anniella pulchra*). Has 30 scale rows anteriorly, 26 medially, and 16 near anus. Shiny, polished, has variable color above, yellow below; faintly lined or unmarked except on tail (which is often heavily pigmented); head dark. Two subspecies: (1) *A. p. pulchra*—silvery, gray, or beige above, with vertebral and several dark lateral lines; (2) *A. p. nigra*—young like (1) but mature individuals black or dark brown above.

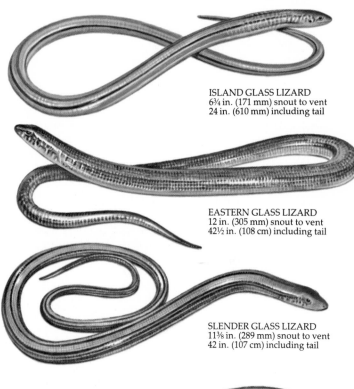

ISLAND GLASS LIZARD
6¾ in. (171 mm) snout to vent
24 in. (610 mm) including tail

EASTERN GLASS LIZARD
12 in. (305 mm) snout to vent
42½ in. (108 cm) including tail

SLENDER GLASS LIZARD
11⅜ in. (289 mm) snout to vent
42 in. (107 cm) including tail

CALIFORNIA LEGLESS LIZARD
6½ in. (165 mm) snout to vent
11¼ in. (286 mm) including tail

VENOMOUS LIZARDS—family Helodermatidae

A single genus (*Heloderma*) and two species comprise this family, the only venomous lizards in the world: the Gila Monster (*H. suspectum*), of the southwestern United States and adjacent Mexico, and the Beaded Lizard (*H. horridum*), of western Mexico. In both species the body is large and stout, the tail short. The back and sides are covered with bony tubercles surrounded by tiny granules; there are smooth, flat, rectangular scales on the belly. A gular fold is present, and loose skin on the neck. The jaws are powerful and all teeth are grooved; venom glands empty into the floor of the mouth. These lizards have no femoral pores, and rarely have a preanal. They are oviparous. Broadly carnivorous, both species are partial to eggs—sometimes the only food acceptable to captives. Their venom is powerful and neurotoxic, but seldom fatal to humans.

 GILA MONSTER (*Heloderma suspectum*). May be yellow, pink, or orange on back and sides. Two subspecies: (1) *H. s. suspectum*—crossbands obliterated by black mottling, dark tail bands and interspaces mottled; (2) *H. s. cinctum*—crossbands in adults as in young (4 on body, 5 on tail), little mottling in dark tail bands.

TEIIDS—family Teiidae

Teiids (40 genera, 225 species) are confined to the Americas. "Macroteiids," an informal group containing all North American species, are diurnal carnivores, terrestrial (a few South American species are semiarboreal or semiaquatic), oviparous, strong-limbed, fast runners; they have femoral pores and large scales on the head. The scales of United States teiids are granular above, quadrangular below. They have a double gular fold, scaly eyelids, conspicuous ear openings, a pointed head, and elongate body and tail. Teiids have a striped, checkered, crossbarred, or spotted (never blotched) pattern.

Whiptails and racerunners (*Cnemidophorus*) are a diverse group, with 15 United States species now recognized. Most live in the dry regions of the Southwest; only one species is found in the eastern United States. *Ameiva*, an introduced genus, is believed to be more primitive. Abundant in the West Indies, and Central and South America, *Ameiva* contains 15–20 species with the same general body form as whiptails but with 10–12 rows of ventral plates (eight in whiptails).

 **GIANT AMEIVA** (*Ameiva ameiva*). Has large central gular scales; 15–23 femoral pores. Some have brown or yellowish vertebral stripe. Introduced subspecies is supposedly *A. a. petersi* of upper Amazon valley, but at least 2 subspecies have been introduced in the Miami area.

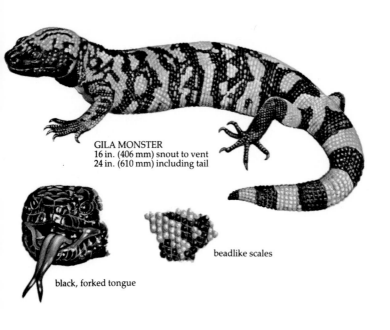

GILA MONSTER
16 in. (406 mm) snout to vent
24 in. (610 mm) including tail

beadlike scales

black, forked tongue

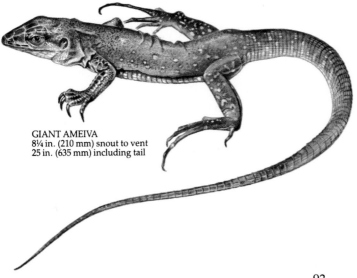

GIANT AMEIVA
8¼ in. (210 mm) snout to vent
25 in. (635 mm) including tail

WHIPTAILS AND RACERUNNERS—genus *Cnemidophorus*

This genus includes 44 species, 15 in the U.S. Nine species are unisexual, reproducing parthenogenetically (without fertilization of eggs). Central gulars are not enlarged; 8 rows of ventrals are present.

KEY TO WHIPTAILS AND RACERUNNERS

1. 1 frontoparietal . . . .
. . . . **Orange Throat**, p. 96
2 frontoparietals . . . . *see* 2
2. (*not shown*) Anterior border of gular fold with abruptly enlarged scales . . . . *see* 5
Small or slightly enlarged scales . .
. . . . *see* 3
3. Supraorbital semicircles to or past middle of 2nd supraocular . . . .
. . . . **New Mexico**, p. 100
Semicircles less extensive . . *see* 4
4. Rear forearm scales granular; pattern variable but rarely as below; 83–100 granules at midbody . . . . **Western**, p. 96
Somewhat enlarged medially; 7 light stripes, unspotted between; 55–71 granules at midbody . . . . **Little Striped**, p. 96
5. (*not shown*) Supraorbital semicircles to or past middle of 2nd supraocular; rear forearm scales moderately enlarged medially .
. . . . . . **Gray-checkered**, p. 100
Semicircles less extensive, or rear forearm scales all granular .*see* 6
6. (*not shown*) No light spots or bars in dark fields between stripes . .
. . . . *see* 7
Spots or bars visible . . . . . . *see* 9
7. Fewer than 6 scales between chin shields and lower labials on either side, or more than 3 preanal scales . . . .
. . . . **Plateau Striped**, p. 98
More than 6 scales, or fewer preanals . . . . *see* 8

8. (*not shown*) Rear forearm scales enlarged, angular medially . . . . .
. . . . **Desert Grassland**, p. 98
Scales granular or slightly enlarged, rounded . . . .
. . **Six-lined Racerunner**, below
9. (*not shown*) Rear forearm scales granular, not enlarged . . . .
. . **Colorado Checkered**, p. 100
Scales at least slightly enlarged medially . . . .*See* 10
10. (*not shown*) 7–8 light stripes or vestiges . . . .*see* 11
6 stripes or vestiges . . . . . .*see* 13
11. (*not shown*) Rear forearm scales weakly enlarged medially . . . .
. . . . **Laredo Striped**, p. 98
Strongly enlarged . . . .*see* 12
12. Males E of 106th meridian with pink throat, dark blue chest, stripes . . . **Texas Spotted**, p. 96
Males W of 106th meridian white below, stripes lost . . . .
. . . . . **Canyon Spotted**, p. 96
13. (*not shown*) 85–115 granules across midbody . . . .
. . . . **Canyon Spotted**, p. 96
65–84 granules . . . .*see* 14
14. (*not shown*) Usually 3–5 granules between paravertebral lines . . .
. . . . **Gila Spotted**, p. 98
Usually 6–8 granules . . . . . .*see* 15
15. (*not shown*) Interparietal width/ length ratio .57–.633 . . . .
. . . **Chihuahuan Spotted**, p. 98
Width/length ratio .659–.955 . . . .
. . . . . . **Sonoran Spotted**, p. 98

Bisexual Species: Six in the U.S., all diploid and reproducing sexually.

SIX-LINED RACERUNNER (*Cnemidophorus sexlineatus*). Light stripes throughout life, never light spots in spaces between. Light blue tail in young. Males suffused with blue below. Two subspecies: (1) *C. s. sexlineatus*—6-lined, no anterior-posterior color gradation; (2) *S. c. viridis*— 7-lined, bright green anteriorly.

KEY TO WHIPTAILS

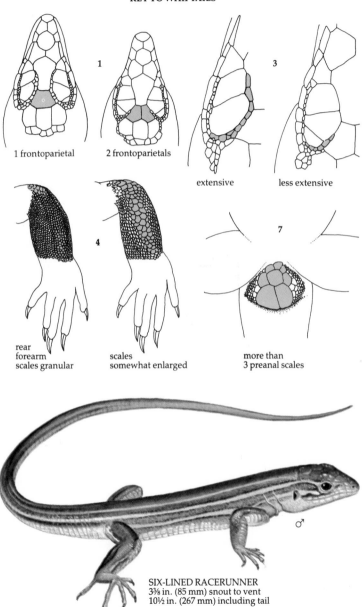

1 frontoparietal

2 frontoparietals

extensive

less extensive

rear
forearm
scales granular

scales
somewhat enlarged

more than
3 preanal scales

♂

SIX-LINED RACERUNNER
3⅜ in. (85 mm) snout to vent
10½ in. (267 mm) including tail

95

LITTLE STRIPED WHIPTAIL (*Cnemidophorus inornatus*). Resembles Six-lined Racerunner, but has 7 stripes. Belly and sides of head are bright blue in males. Two subspecies in U.S.: (1) *C. i. arizonae*—blue-gray above, light blue below; (2) *C. i. heptagrammus*—gray-black to black above, blue below.

TEXAS SPOTTED WHIPTAIL (*Cnemidophorus gularis*). Has light stripes throughout life; except in very young, the spaces between stripes bear light dots, at least on sides. Two subspecies in U.S.: (1) *C. g. gularis*—7–8 lines extending to tail with round dots between, chest dark blue, throat pink in males; (2) *C. g. septemvittatus*—6–7 lines to tail with dots between transversely expanded at least on sides, lines faded posteriorly, chest dark-flecked, throat whitish in males.

CANYON SPOTTED WHIPTAIL (*Cnemidophorus burti*). Has 6 (rarely 7) light lines, disappearing or becoming faint in large adults. Young have red tail. Two subspecies in U.S.: (1) *C. b. stictogrammus*—only head and neck reddish above; (2) *C. b. xanthonotus*—all dorsal surfaces reddish.

WESTERN WHIPTAIL (*Cnemidophorus tigris*). Highly variable in color and pattern, from uniformly brownish gray to mottled, checkered, marbled above and white to spotted or black below (often just chest and throat). Tail is bright blue in young, gray in adults.

Six subspecies in U.S.: (1) *C. t. tigris*—4 dorsal light lines, sides barred, dark-spotted or not below; (2) *C. t. mundus*—8 light lines, sides not barred, throat with few dark spots; (3) *C. t. stejnegeri*—like (2) but lateral lines dim, throat heavily spotted; (4) *C. t. gracilis*—4 dorsal light brown lines, sides spotted, suffused with black below; (5) *C. t. septentrionalis*—like (4) but light lines yellow and not reaching tail, throat with small black spots; (6) *C. t. marmoratus*—light lines scarcely evident, mottled above, sides barred, throat and chest spotted.

ORANGE THROAT WHIPTAIL (*Cnemidophorus hyperythrus*). Six-lined, with paravertebrals weakest, laterals brightest, spaces between unmarked, darker laterally. Belly is bluish white; throat is orange; tail is blue in young. One subspecies in U.S.: *C. h. beldingi*.

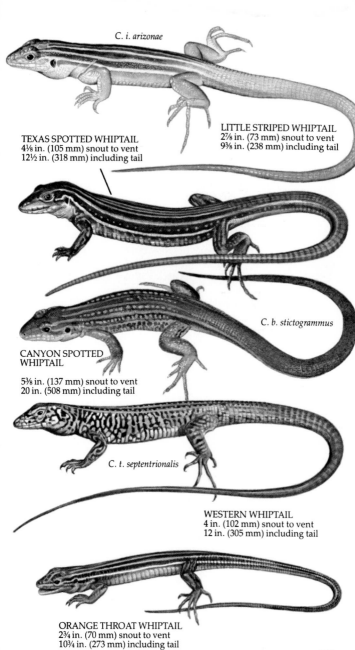

C. i. arizonae

LITTLE STRIPED WHIPTAIL
2⅞ in. (73 mm) snout to vent
9⅜ in. (238 mm) including tail

TEXAS SPOTTED WHIPTAIL
4⅛ in. (105 mm) snout to vent
12½ in. (318 mm) including tail

C. b. stictogrammus

CANYON SPOTTED
WHIPTAIL

5⅜ in. (137 mm) snout to vent
20 in. (508 mm) including tail

C. t. septentrionalis

WESTERN WHIPTAIL
4 in. (102 mm) snout to vent
12 in. (305 mm) including tail

ORANGE THROAT WHIPTAIL
2¾ in. (70 mm) snout to vent
10¾ in. (273 mm) including tail

Unisexual Species: Males do not occur (or only rarely) in the following nine species of whiptails. Females reproduce parthenogenetically (without fertilization of eggs). All species arose by hybridization of bisexual species. Three clear-cut pattern groups are recognized: (1) striped and unspotted (Plateau Striped, Desert Grassland); (2) striped and spotted (Chihuahuan Spotted, Gila Spotted, Sonoran Spotted, Laredo Striped, New Mexico); (3) striped and checkered (Colorado Checkered, Gray-checkered). Separating the species into pattern groups is often difficult; locality, if known, may be helpful.

PLATEAU STRIPED WHIPTAIL (*Cnemidophorus velox*). Has 6–7 light stripes; interspaces black or dark brown with no light marks. White or tinged with bluish green below. Tail blue.

DESERT GRASSLAND WHIPTAIL (*Cnemidophorus uniparens*). Usually has 7 (sometimes 6) yellowish to whitish stripes on body, vertebral often confined to neck; interspaces reddish brown to black. White below; adults often bluish on throat and neck. Tail olive-green to bluish green.

CHIHUAHUAN SPOTTED WHIPTAIL (*Cnemidophorus exsanguis*). Has 6 pale brownish stripes, with vertebral visible on young only; interspaces are brown, sometimes reddish; numerous yellowish-white round dots are scattered over interspaces and stripes. Whitish, faintly suffused with blue, below. Tail dark, often faintly greenish. Young have blue-green tail, dots pale reddish. Vicious in captivity, often killing others of same species.

GILA SPOTTED WHIPTAIL (*Cnemidophorus flagellicaudus*). Like Chihuahuan Spotted but fewer light dots, paravertebral lines close together, lines sharper; 3½ in. (89 mm) snout to vent, 12 in. (305 mm) including tail. Young without light dots. [Not illustrated.]

SONORAN SPOTTED WHIPTAIL (*Cnemidophorus sonorae*). Closely resembles Chihuahuan Spotted but light lines more sharply defined. Young lack light dots.

LAREDO STRIPED WHIPTAIL (*Cnemidophorus laredoensis*). Has 7 cream or white lines, vertebral as distinct as others; spaces between are dark green or greenish brown; light dots on interspaces at sides on rear. Tail greenish brown above, tan below; undersides white. Believed to be hybrid of Six-lined and Texas Spotted.

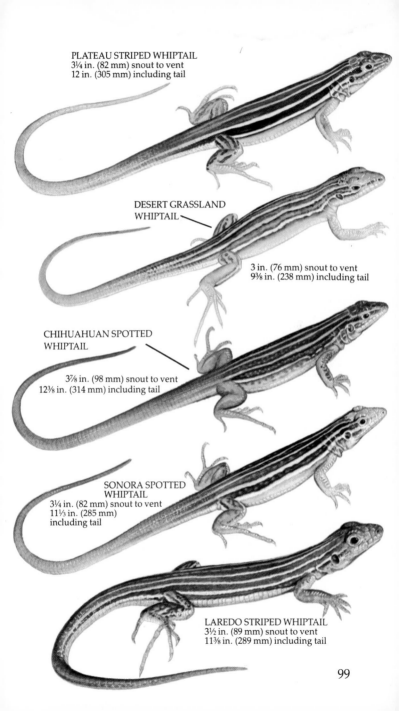

PLATEAU STRIPED WHIPTAIL
3¼ in. (82 mm) snout to vent
12 in. (305 mm) including tail

DESERT GRASSLAND WHIPTAIL
3 in. (76 mm) snout to vent
9⅜ in. (238 mm) including tail

CHIHUAHUAN SPOTTED WHIPTAIL
3⅞ in. (98 mm) snout to vent
12⅜ in. (314 mm) including tail

SONORA SPOTTED WHIPTAIL
3¼ in. (82 mm) snout to vent
11⅓ in. (285 mm) including tail

LAREDO STRIPED WHIPTAIL
3½ in. (89 mm) snout to vent
11⅜ in. (289 mm) including tail

99

NEW MEXICO WHIPTAIL (*Cnemidophorus neomexicanus*). Resembles Laredo Striped but vertebral light lines are wavy, spaces between brown to black, tail greenish gray (blue in young). Believed to be a hybrid of Western and Little Striped.

COLORADO CHECKERED WHIPTAIL (*Cnemidophorus tesselatus*). Hybrid of Western, Little Striped, and Six-lined. Considerable pattern variation. Has 6–8 stripes, vertebral double, single, or lacking. Light dots or bars break dark fields into rectangular spots that appear checkered. Thorax and neck black-flecked in some.

GRAY-CHECKERED WHIPTAIL (*Cnemidophorus dixoni*). Like Colorado Checkered but stripes more numerous (8–10), checks smaller and more numerous; 3⅞ in. (98 mm) snout to vent, 14⅜ in. (365 mm) including tail. [Not illustrated.]

IGUANID LIZARDS—family Iguanidae

Iguanids (8 subfamilies, 65 genera, 650 species) are limited to the Western Hemisphere except for 3 genera of Pacific and Indian ocean islands. Iguanids include 14 of the 27 genera of lizards native to the U.S. In most species, mature males have well-developed femoral pores, rudimentary but visible in females and young.

KEY TO IGUANIDS

1. Digits partly padlike . . . .
 . . . .**anoles,** p. 102
 Digits not expanded . . . . . . *see* **2**
2. Spiny horns at rear of head . . . . .
 . . . . . .**horned lizards,** p. 124
 No horns or spines . . . . *see* **3**
3. Scales at upper edge of orbit overlapping . . . . *see* **4**
 No overlap . **Chuckwalla,** p. 106
4. Lip scales overlapping . . . . *see* **5**
 Not overlapping . . . . *see* **8**
5. Ear openings present . . . . . *see* **7**
 No ear openings . . . . *see* **6**
6. Broad black bands under tail . . . .
 . . . . . . **Greater Earless,** p. 130
 Narrow or none . . . .
 . . . .**earless lizards,** p. 128
7. Interparietal smaller than ear opening . . . .
 . . . .**fringe-toed lizards,** p. 130
 Larger . . **Zebratail Lizard,** p. 130
8. Throat fold with granules . .*see* **10**
 No fold with granules . . . . . *see* **9**
9. Median dorsal row of enlarged granules . . . . . .**Curltail,** p. 132
 Not so . . . . . **spiny lizards,** p. 114
10. Median row of enlarged scales . . .
 . . . .*see* **11**
 Not as above . . . .*see* **12**
11. Whorls of enlarged spines on tail . . . . . .**Short-crested,** p. 104
 Not so . . . . **Desert,** p. 104
12. Interparietal smaller than ear opening . . . .*see* **13**
 Larger . . . .*see* **14**
13. 1–2 rows of large scales between orbits . . **collared lizards,** p. 106
 Several rows of small scales between . . **leopard lizards,** p. 108
14. No enlarged dorsals . . . .
 . . **Banded Rock Lizard,** p. 110
 Some or all enlarged . . . . . .*see* **15**
15. Dorsolateral fold with enlarged scales . . . . . **tree lizards,** p. 112
 No folds . . **Side-blotched,** p. 110

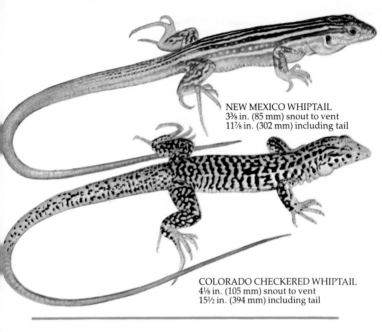

NEW MEXICO WHIPTAIL
3⅜ in. (85 mm) snout to vent
11⅞ in. (302 mm) including tail

COLORADO CHECKERED WHIPTAIL
4⅛ in. (105 mm) snout to vent
15½ in. (394 mm) including tail

KEY TO IGUANIDS

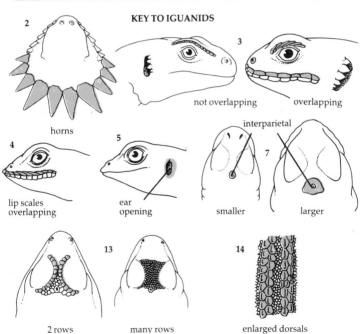

2 horns

3 not overlapping   overlapping

4 lip scales overlapping

5 ear opening

7 interparietal   smaller   larger

13 2 rows   many rows

14 enlarged dorsals

ANOLINE LIZARDS—subfamily Anolinae

Anolis, the only United States genus of the 15 in this subfamily, has more species (over 300) than any other genus of lizards. All lack femoral pores; most have slightly enlarged or no postanals. Digits are expanded (with a few exceptions in the tropics). Anoles lay one egg every two weeks over an eight-month period. The throat fan is well developed in males, small or lacking in females. Most are climbers. All change colors rapidly from brown to green or light to dark, and may blend almost indistinguishably with their background. Only one species, the Green Anole, is native to the United States.

KEY TO ANOLES

1. Belly scales keeled . . . . *see* **2**
   Smooth . . . . *see* **3**
2. Tail round . . . . . . **Green,** below
   Strongly compressed . . . .
   . . . . **Brown,** p. 104
3. Single median row of spinelike scales on back . . **Knight,** p. 104
   Not as above . . . . *see* **4**
4. Interparietal larger than ear opening . . . . **Bark,** below
   Smaller . . . . . **Bigheaded,** below

GREEN ANOLE (*Anolis carolinensis*). Has keeled scales, with middorsals slightly enlarged. Tail round or oval in cross section; 3 middorsal scales per whorl. Dewlap usually pink, sometimes gray or bluish. Postanals are enlarged in males. One N.A. subspecies: *A. c. carolinensis*—native to Gulf states.

BIGHEADED ANOLE (*Anolis cybotes*). Has minute granular scales, tubercular but not keeled. Vertebrals slightly enlarged; ventrals smooth. Tail is flattened, with serrate ridge; whorls are as in Green. Dewlap gray to orange-yellow; postanals slightly enlarged; erectile median fold on neck and back. Head is enlarged in males. Never all green. Females have straight-edged or scalloped light median stripe. One U.S. subspecies: *A. c. cybotes*—from Hispaniola.

BARK ANOLE (*Anolis distichus*). Scales are minute granules, not enlarged at midline; ventrals smooth. Tail slightly compressed; 5 middorsal scales per whorl. Median scales on snout are mostly paired. In males, dewlap is yellow to orangish; postanals slightly enlarged. Green or mottled, usually has dark band between eyes; tail is banded; often has narrow diagonal crossbands on back and limbs. Two subspecies introduced: *A. d. dominicensis*—circumorbital scale rows contacting, often green, from Hispaniola; *A. d. floridanus*—rows separated, never green, probably from Andros Island, Bahamas.

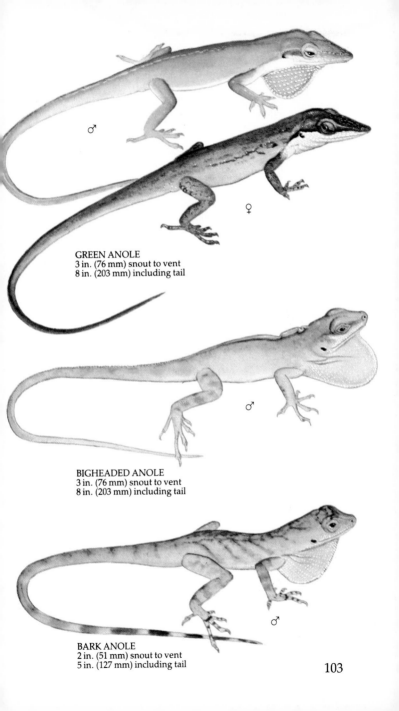

GREEN ANOLE
3 in. (76 mm) snout to vent
8 in. (203 mm) including tail

BIGHEADED ANOLE
3 in. (76 mm) snout to vent
8 in. (203 mm) including tail

BARK ANOLE
2 in. (51 mm) snout to vent
5 in. (127 mm) including tail

103

KNIGHT ANOLE (*Anolis equestris*). Has flat-topped helmetlike head with enlarged tubercles at edges. Dorsal and lateral scales are small and flat, with median row enlarged and spinelike; ventrals are smooth. In males, dewlap is yellow to pink, and there are no enlarged postanals. Tail is flattened, with crest. Usually green, sometimes brown. One subspecies in U.S.: *A. e. equestris*—from Cuba.

BROWN ANOLE (*Anolis sagrei*). All scales are keeled. Tail is strongly compressed, with whorls as in Bark. In males, dewlap is orange-red with yellow or white margins, and postanals are slightly enlarged. Never green. Females and young have light streak or series of rhombs; males have light streak on throat and chest, dark-bordered when dewlap is closed. Two subspecies introduced: *A. s. sagrei*—circumorbital scale rows not in contact, from Cuba; *A. s. ordinatus*—circumorbitals in contact, from Bahamas.

IGUANINE LIZARDS—subfamily Iguaninae

The nine genera of this primitive subfamily have only two native representatives in the United States: *Dipsosaurus* and *Sauromalus*. *Ctenosaura* may have occurred naturally at one time in Arizona (*C. hemilopha*) or Texas (*C. acanthura*). These genera have tiny scales, lack enlarged postanals in males, and are completely or strongly herbivorous (all other iguanids are predominantly carnivorous).

SHORT-CRESTED SPINY-TAILED IGUANA (*Ctenosaura hemilopha*). Introduced from Southern Baja California or NW Mexico into vicinity of Fullerton, California. Has vertebral row of spines, rings of spines on tail. Subspecies in the U.S. uncertain. [No range map.]

Similar, but with crest complete to tail, is the Broad-ringed Spiny-tailed Iguana (*C. pectinata*), introduced near Brownsville, Texas, from W Mexico (Colima).

DESERT IGUANA (*Dipsosaurus dorsalis*). Has minute, keeled dorsal scales, scarcely larger on head but enlarged in middorsal row; belly scales are small, smooth. There are 18–26 femoral pores per side. Reddish brown to gray above, whitish below; throat is lightly mottled. Numerous small round light spots partially dark-margined on back; short longitudinal dark streaks on sides; many narrow crossbars on tail. One subspecies in U.S.: *D. d. dorsalis*.

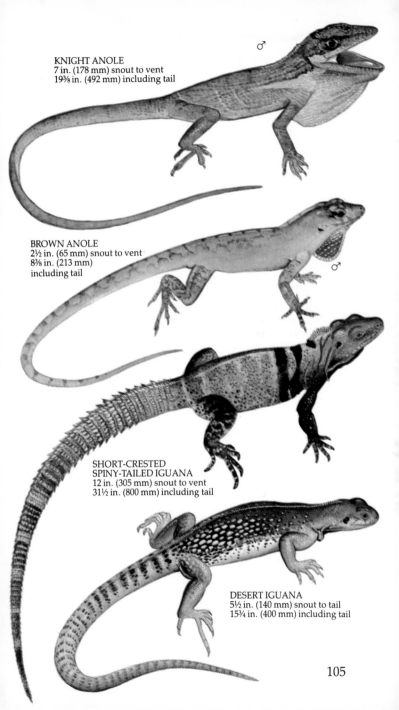

KNIGHT ANOLE
7 in. (178 mm) snout to vent
19⅜ in. (492 mm) including tail

♂

BROWN ANOLE
2½ in. (65 mm) snout to vent
8⅜ in. (213 mm)
including tail

♂

SHORT-CRESTED
SPINY-TAILED IGUANA
12 in. (305 mm) snout to vent
31½ in. (800 mm) including tail

DESERT IGUANA
5½ in. (140 mm) snout to tail
15¾ in. (400 mm) including tail

105

CHUCKWALLA (*Sauromalus obesus*). The largest native iguanid. Scales on body are small; somewhat larger on limbs, tail, and head. Also has prominent throat fold. There are 11–24 femoral pores per side. Young have 4–5 broad brown crossbands on gray-brown background; 3–6 on tail. Bands disappear in adults, which are irregularly washed with red, yellow, brown, and slate above and below. Three subspecies in U.S.: (1) *S. o. obesus*—scales small, over 50 encircling middle of forearm; (2) *S. o. tumidus*—scales larger, fewer than 50; (3) *S. o. multiforaminatus*—double rows of femoral pores.

CROTAPHYTINE LIZARDS—subfamily Crotaphytinae

This subfamily is comprised of only two genera, *Crotaphytus* and *Gambelia*. Their body scales are small; limbs and tail are long and strong. They are bipedal when running fast—a rarity in lizards—and are strongly predatory.

COLLARED LIZARDS—genus *Crotaphytus*

The head is exceptionally broad, the mouth large, and the neck narrow in these lizards. Femoral pores number 14–27. Males lack enlarged postanals. Tail scales are distinctly larger than the small, smooth body scales; the dorsal scales on the head are somewhat enlarged. Eggs number 4–24 (seldom fewer than 8), and are laid in sandy soils or burrows, or under rocks.

KEY TO COLLARED LIZARDS

1. No black collars, or 1 vertical bar at sides; no postfemoral pocket . . . . **Reticulate**, p. 108
1 or 2 collars, pocket . . . . . . *see* **2**

2. Inner lining of throat white; tail compressed laterally . . . . **Black-collared**, p. 108
Throat lining black; tail round . . . . **Collared**, below

COLLARED LIZARD (*Crotaphytus collaris*). Has 2 black collars. Five subspecies in U.S.: (1) *C. c. collaris*—circumorbital rows fused at 1 or more scales medially, gular region not reticulated (it is in all others); (2) *C. c. baileyi*—back green or blue, head not yellow or, if so, not on chin or back of orbits; (3) *C. c. auriceps*—back green or blue, head yellow under chin as well as back to rear collar; (4) *C. c. fuscus*—back brown, never green or blue, head not yellow, light dorsal spots the size of lateral ones; (5) *C. c. nebrius*—no green or blue coloration, light dorsal spots 1½–3 times larger than lateral spots.

CHUCKWALLA
8¼ in. (210 mm) snout to vent
16½ in. (419 mm) including tail

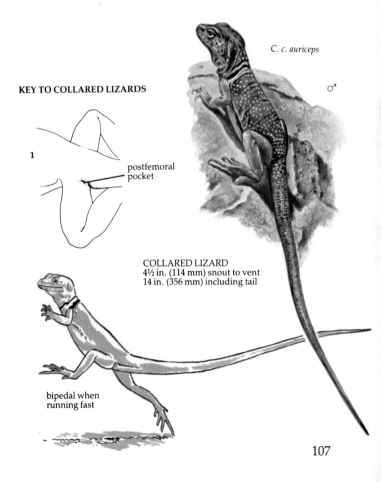

C. c. auriceps

♂

KEY TO COLLARED LIZARDS

1

postfemoral
pocket

COLLARED LIZARD
4½ in. (114 mm) snout to vent
14 in. (356 mm) including tail

bipedal when
running fast

107

BLACK-COLLARED LIZARD (*Crotaphytus insularis*). Has black in groin and gular area in males. Two subspecies in U.S: (1) *C. i. vestigium*—rear collar reduced or lacking, anterior collar broadly interrupted (16 or more scales between ends), with no dark spots; (2) *C. i. bicinctores* (now considered a full species, Mojave Black-collared Lizard)—anterior collar interrupted by no more than 12 scales.

RETICULATE COLLARED LIZARD (*Crotaphytus reticulatus*). Lacks black collars or has only 1 vertical bar on sides of neck. Coarse network of narrow light lines on top and sides of head, body, and limbs encloses numerous eye-sized areas of gray-brown to reddish brown. Some ocelli in the 4 rows on back and sides are black. Gular region and femoral pores are black in males. Dim network on chin and sides of jaw and abdomen, otherwise whitish below. Hatchlings are light gray, with 4–6 yellow to yellow-orange crossbands separating rows of black spots.

LEOPARD LIZARDS—genus *Gambelia*

These lizards have a relatively long and slender head. Males have enlarged postanals. Femoral pores number 14–25. Head scales are not particularly enlarged except for a subocular and a few canthals. Tail scales are scarcely larger than the minute, smooth body scales. There are 2–9 eggs per clutch, usually 7 or fewer.

BLUNTNOSE LEOPARD LIZARD (*Gambelia silus*). Has shorter, broader head than Longnose, width equal to or greater than distance from nostril to anterior border of ear. Throat mottled or blotched with dark. Gray or brown above, with 7–10 broad light bars extending onto sides; small, indistinct reddish spots between bars.

LONGNOSE LEOPARD LIZARD (*Gambelia wislizenii*). Has longer, narrower head than Bluntnose, width less than distance from nostril to anterior border of ear. Longitudinal dark streaks on throat. Light bars are as in Bluntnose, but do not extend onto sides and have dark spots between; bars become narrow, indistinct, or lacking in adults. Three N.A. subspecies: (1) *G. w. wislizenii*—large dark spots undivided, bordered by light dots; (2) *G. w. punctatus*—irregular small dark spots, not surrounded by light dots; (3) *G. w. copei*—large dark spots divided, each part surrounded by light dots.

BLACK-COLLARED LIZARD
5½ in. (140 mm) snout to vent

C. i. bicinctores

♂

RETICULATE COLLARED LIZARD
5⅜ in. (137 mm) snout to vent
16¾ in. (425 mm) including tail

BLUNTNOSE LEOPARD LIZARD
4½ in. (114 mm) snout to vent
12½ in. (318 mm) including tail

♀

LONGNOSE LEOPARD LIZARD
5 in. (127 mm) snout to vent
15⅛ in. (384 mm) including tail

SCELOPORINE LIZARDS—subfamily Sceloporinae

Most iguanid lizards in North America belong to this subfamily and fit into four groups: (1) the most primitive—banded rock lizards, *Streptosaurus*; (2) scramblers—*Uta*, *Urosaurus*, and *Sceloporus*, derived from the first group; (3) highly specialized horned lizards—*Phrynosoma*; and (4) highly specialized sand lizards—*Holbrookia*, *Cophosaurus*, *Callisaurus*, and *Uma*. Sceloporines have femoral pores and enlarged head scales, including large interparietal. Iguanines and crotaphytines have femoral pores but mostly small head scales, tiny interparietal. Anolines and tropidurines lack femoral pores and have large interparietal. Males have enlarged postanals except in *Phrynosoma*.

 BANDED ROCK LIZARD (*Streptosaurus mearnsi*). Often placed in genus *Petrosaurus*, this lizard is conspicuously flat-bodied. Dorsal scales granular; ventrals slightly larger, overlapping, and smooth. Granular gular fold. Tail scales large, strongly keeled, and pointed. Head scales irregularly enlarged. Femoral pores, 19–25. Shallow postfemoral skin pocket. Single bright black collar. In males, light bluish dots below, spots on throat pinkish or white. Gravid females have orange on throat and above eye. One subspecies in U.S.: *S. m. mearnsi*.

 SIDE-BLOTCHED LIZARD (*Uta stansburiana*). Most common lizard of SW. Dorsal scales are small but overlapping, keeled, larger on tail. Ventrals larger, smooth. Head scales irregularly enlarged. Granular gular fold. Femoral pores, 11–17. Color and pattern are highly variable. In some, dorsolateral light stripes enclose area with 7–9 pairs of dark spots; similar spots on sides. Stripes and spots disappear or become dim when lizard matures, often replaced by light blue dots. Black patch behind axilla (lacking in females of some subspecies) is basis for common name.

Six subspecies in U.S.: (1) *U. s. stansburiana*—patterned but light stripes weak or absent, as are dark spots in males; (2) *U. s. elegans*—stripes and spots distinct; (3) *U. s. hesperis*—2 postrostrals between internasal and rostral (1 in others); (4) *U. s. stejnegeri*—fewer than 8 scales between femoral pore series (more in others); (5) *U. s. nevadensis*—adults mostly unicolor, males orange at sides of belly in reproductive season, "bobs" once in display or alarm; (6) *U. s. uniformis*—adults unicolor, males not orange, 4–6 "bobs."

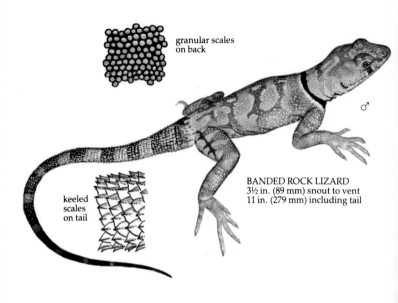

granular scales on back

keeled scales on tail

BANDED ROCK LIZARD
3½ in. (89 mm) snout to vent
11 in. (279 mm) including tail

♂

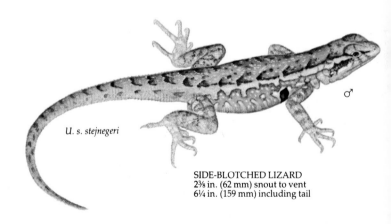

U. s. stejnegeri

♂

SIDE-BLOTCHED LIZARD
2⅜ in. (62 mm) snout to vent
6¼ in. (159 mm) including tail

TREE LIZARDS—genus *Urosaurus*

Small, amazingly camouflaged, quick-moving, and elusive, these lizards "bob" frequently. They are diurnal, egg laying, and insectivorous. Ear openings are present, and one or more lateral skin folds with enlarged scales. Also present is a granular gular fold; median dorsals are enlarged.

KEY TO TREE LIZARDS

1. Postfemoral skin pocket . . . *see* 2
   No skin pocket . . . . **Brush**, below

2. Vertebral scales smaller than enlarged adjacents . . **Tree**, below
   Same size as adjacents . . . .
   . . . . **Small-scaled**, below

BRUSH LIZARD (*Urosaurus graciosus*). Has a long tail, at least twice the head-body length; central strip of enlarged scales; 9–14 femoral pores. Males have dotted blue patch on sides of belly. Two subspecies: (1) *U. g. graciosus*—pattern faint; (2) *U. g. shannoni*—pattern bold.

TREE LIZARD (*Urosaurus ornatus*). Has mostly granular dorsal scales, 1 or 2 rows on midline enlarged and bordered by a few rows much enlarged and keeled. Femoral pores, 10–16. Collarlike crossbar at shoulders or neck. In males, sides or entire belly blue; throat blue, orange, or yellow. Six subspecies in U.S.: (1) *U. o. ornatus*—2 series of enlarged paravertebral scales, irregularly arranged, inner almost twice the size of outer; (2) *U. o. schmidti*—like (1) but inner series not twice the size of outer; (3) *U. o. symmetricus*—2 paravertebral series, regularly arranged, separated by vertebral areas as wide as widest dorsal; (4) *U. o. schotti*—like (3) but vertebral area narrower; (5) *U. o. levis*—3 or 4 paravertebral series, enlarged dorsals at base of tail; (6) *U. o. wrighti*—like (5) but scales at base of tail not enlarged.

SMALL-SCALED LIZARD (*Urosaurus nigricaudus*). Has minute scales, a few middorsal rows slightly enlarged. Frontal is entire (divided in other N.A. tree lizards). There are 10–15 femoral pores. Males have yellow to orange throat, blue on sides of belly, light to dark gray below. Females and young lack blue belly and yellow-orange colors. One subspecies in U.S.: *U. n. microscutatus*.

112

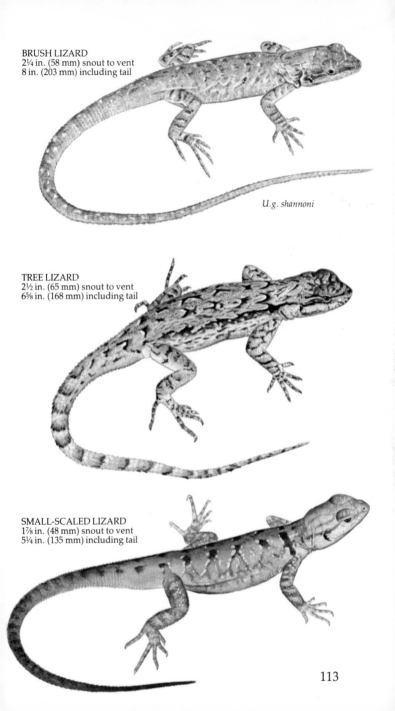

BRUSH LIZARD
2¼ in. (58 mm) snout to vent
8 in. (203 mm) including tail

U.g. shannoni

TREE LIZARD
2½ in. (65 mm) snout to vent
6⅝ in. (168 mm) including tail

SMALL-SCALED LIZARD
1⅞ in. (48 mm) snout to vent
5¼ in. (135 mm) including tail

113

SPINY LIZARDS—genus *Sceloporus*

In number of species (16), this is the largest genus of lizards in the United States. All have rather large keeled, spiny scales on body and tail, enlarged scales on head, femoral pores, enlarged postanals in males, no granular throat fold, and, with one exception, brightly colored areas on the sides of the belly in males. Spiny lizards are carnivorous. *S. cyanogenys, grammicus, jarrovi,* and *poinsetti* are viviparous; all others are oviparous.

KEY TO SPINY LIZARDS

1. (*not shown*) Broad light-bordered black collar . . . . *see* 2
   Not as above . . . . *see* 4
2. (*not shown*) Enlarged supraoculars in 1 row . . . . . **Yarrow's**, p. 118
   In 2 rows or irregular . . . . . *see* 3
3. (*not shown*) Bright bands encircling tail . . . . **Crevice**, p. 118
   Not encircling, dimmer . . . . **Blue**, p. 118
4. Supraoculars large, rear 1 or 2 contacting median head scales . . . . *see* 5
   Smaller, all bordered medially by row of small scales . . . . . *see* 7
5. 1st lower labial separated from 1st postmental . . . . **Desert**, p. 120
   Contacting 1st postmental . . *see* 6
6. Upper ear scales largest . . . . **Clark's**, p. 120
   Median scales largest . . . . **Granite**, p. 120
7. Postfemoral pocket present . . . . **Rosebelly**, below
   No postfemoral pocket . . . . *see* 8
8. Lateral scales in straight rows . . . . **Bunch Grass**, p. 116
   In oblique rows . . . . *see* 9
9. Lateral scales granular, not overlapping . . . . . . **Canyon**, p. 116
   Overlapping . . . . *see* 10

10. (*not shown*) Scales on rear of thigh granular, not overlapping *see* 11
    Overlapping . . . . *see* 12
11. (*not shown*) Lateral and dorsal scales on neck differing abruptly in size . . . . **Mesquite**, p. 116
    Blending in size . . . . **Sagebrush**, p. 120
12. (*not shown*) 1 row of enlarged supraoculars covering three-fourths or more of supraocular area . . . . **Texas**, p. 118
    Covering two-thirds or less . *see* 13
13. (*not shown*) Scales in center of rear surface of thigh abruptly smaller than dorsal scales, not much smaller to sides . . . . **Western Fence**, p. 122
    Blending in size with dorsals . . . . *see* 14
14. Found in Florida . . . . *see* 15
    Found outside Florida . . . . *see* 16
15. (*not shown*) Dorsals smaller, 38 or more from head to base of tail . . . . **Florida Scrub**, p. 122
    Larger, 37 or fewer . . . . . . *see* 16
16. (*not shown*) Males unmarked below . . . **Striped Plateau**, p. 122
    Males blue-black on sides of belly, often on throat . . . . **Eastern Fence**, p. 122

Miscellaneous Group: All these spiny lizards are small or medium in size and regarded as more primitive than the other groups.

ROSEBELLY LIZARD (*Sceloporus variabilis*). Small dorsal scales, 58–69 from head to above anus; 10–14 femoral pores; postfemoral skin pocket. In males, sides are dark below dorsolateral line; black slash in axilla, continuous with dark borders of pink lateral belly patches. One U.S. subspecies: *S. v. marmoratus.*

114

KEY TO SPINY LIZARDS

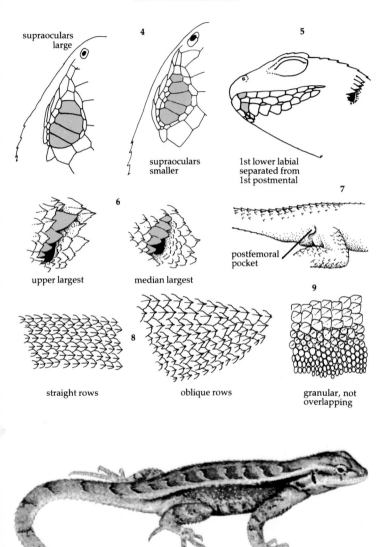

4 supraoculars large

supraoculars smaller

5 1st lower labial separated from 1st postmental

6 upper largest

median largest

7 postfemoral pocket

8 straight rows

oblique rows

9 granular, not overlapping

ROSEBELLY LIZARD
2⅛ in. (54 mm) snout to vent
5½ in. (140 mm) including tail

CANYON LIZARD (*Sceloporus merriami*). Has small dorsal scales, 42–70 from head to above anus; lateral scales granular. Femoral pores number 19–32; the 2 series in contact or separated by up to 7 scales (usually no more than 4); no postfemoral skin pocket. Light gray to blue-gray above, with 10–11 darker, dimly outlined spots on each side of midline, better defined to rear; numerous whitish or pale blue flecks; dark spot in front of arm. In males, sides of belly are lavender bordered with dark blue. Females have dark blue slash at rear of abdomen, otherwise whitish below. Both sexes have bars or 2–3 pairs of dark spots on throat, rear ones largest. Three subspecies in U.S.: (1) *S. m. merriami*—1st infralabial completely separated from postmental (not in others); (2) *S. m. longipunctatus*—paravertebral dark spots comma-shaped, extending laterally; (3) *S. m. annulatus*—spots not extended.

BUNCH GRASS LIZARD (*Sceloporus scalaris*). Has 37–46 dorsal scales from head to above anus. Laterals are same size, in rows mostly parallel with dorsals. Femoral pores number 12–18, the 2 series in contact or separated by no more than 2 scales; 2 postrostrals (4 in most other species); no postfemoral skin pocket. Two color phases—completely dark brown to yellow-brown above, or a bright pattern of stripes and spots. In latter, there is a bright, sharp-edged, narrow dorsolateral light stripe and similar lateral line; 12 dorsal pairs of crescentic dark spots; lateral series of similar marks; blue-centered dark spot in front of arm. Males have blue patch from axilla to groin. Both sexes are otherwise whitish and unmarked below. One subspecies in U.S.: *S. s. slevini*.

MESQUITE LIZARD (*Sceloporus grammicus*). Has small dorsal scales, 52–74 from head to above anus. Lateral scales on neck abruptly smaller than dorsals; 12–20 femoral pores; no postfemoral skin pocket. Well camouflaged for life on trees. Females and young have 2 narrow black paravertebral lines on nape; 4–5 narrow undulate bars across back, 1 collarlike on neck. Limbs and tail are similarly banded. Males lack bars except 1 on neck; center of throat is flesh to pale blue in color; sides of belly pale blue; black flecks on throat and parts of belly. One subspecies in U.S.: *S. g. disparilis*.

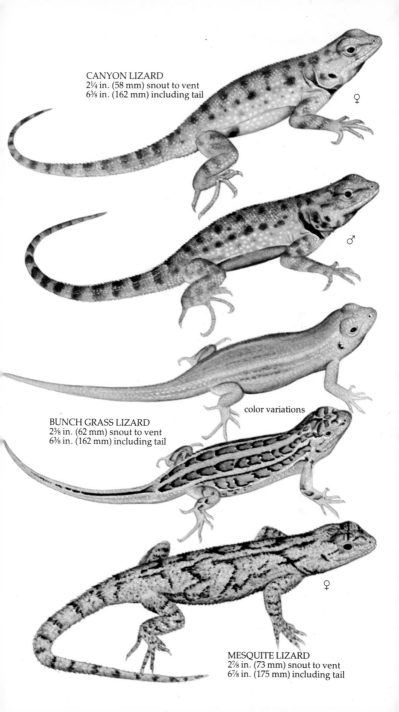

CANYON LIZARD
2¼ in. (58 mm) snout to vent
6⅜ in. (162 mm) including tail

♀

♂

BUNCH GRASS LIZARD
2⅜ in. (62 mm) snout to vent
6⅜ in. (162 mm) including tail

color variations

♀

MESQUITE LIZARD
2⅞ in. (73 mm) snout to vent
6⅞ in. (175 mm) including tail

Collared Group: The 11 species in this group of spiny lizards are medium-large to large, viviparous, and have a broad black collar, light-edged front and rear but never broken medially in United States species. Three species in the United States.

BLUE SPINY LIZARD (*Sceloporus cyanogenys*). Has large dorsal scales, 32–40 from head to above anus, weakly keeled, strongly pointed. Supraoculars large with 1 or more split on each side; 12–17 femoral pores. Largest males are brilliant greenish blue above except for somewhat brownish head and a few light scales on back; throat and chin are pale blue, with light blue patches edged with dark blue on sides of belly. Tail bands are not sharply defined and do not encircle tail.

YARROW'S SPINY LIZARD (*Sceloporus jarrovi*). Males are black above, every scale except on head and neck with light blue center; tail bands are obscure; throat and sides of abdomen are dark blue, belly light blue or gray, chest and groin gray or black. Dorsal scales number 38–46 from head to above anus; 1 series of supraoculars is enlarged. Femoral pores number 13–18. One subspecies in U.S.: *S. j. jarrovi.*

CREVICE SPINY LIZARD (*Sceloporus poinsetti*). Has 31–41 dorsal scales from head to above anus, nearly smooth but with 1 or more small, terminal points. Enlarged supraoculars are in 2 complete rows. Femoral pores number 9–18. Brightly banded tail. One subspecies in U.S.: *S. p. poinsetti.*

Spinose Group: The eight medium-large species in this group have large, strongly pointed scales and one row of greatly enlarged supraoculars, reducing small bordering scales to partial rows except in Texas Spiny. All are oviparous. A black slash may be present in the front of the arm and may extend dorsally as a narrow, interrupted collar. Four species are found in the United States.

TEXAS SPINY LIZARD (*Sceloporus olivaceus*). Has large, rather strongly keeled and pointed dorsal scales, 28–33 from head to anus. Supraoculars are large, 1 series of 5 or 6 scales but none contacting median head scales. Femoral pores, 11–16. One subspecies in U.S.: *S. o. olivaceus.*

BLUE SPINY LIZARD
5¾ in. (146 mm) snout to vent
14⅛ in. (359 mm) including tail

YARROW'S SPINY LIZARD
3½ in. (89 mm) snout to vent
8⅞ in. (225 mm) including tail

CREVICE SPINY LIZARD
4⅝ in. (118 mm) snout to vent
12¼ in. (311 mm) including tail

TEXAS SPINY LIZARD
4¾ in. (121 mm) snout to vent
11 in. (279 mm) including tail

GRANITE SPINY LIZARD (*Sceloporus orcutti*). Large dorsal scales, 29–36 from head to above anus, strongly keeled and pointed, with deep notches and accessory points. Supraoculars very large, with rear 2 contacting median head scales. Femoral pores, 10–15. First lower labial contacts postmental.

CLARK'S SPINY LIZARD (*Sceloporus clarki*). Has large keeled, pointed dorsal scales, 28–36 from head to above anus. Supraoculars very large, some contacting median head scales. Three large scales, upper largest, overlap ears. First lower labial contacts postmental. Femoral pores, 10–16. Two subspecies in U.S.: (1) *S. c. clarki*—juvenile pattern of 6–7 dark undulating crossbars lost in adult males; (2) *S. c. vallaris*—juvenile pattern retained.

DESERT SPINY LIZARD (*Sceloporus magister*). Has large keeled, pointed dorsal scales, 26–37 from head to above anus. Supraoculars very large, some in contact with median head scales. There are 5–7 large scales overlapping ears, largest in middle of series. First lower labial is completely separated from postmental. Femoral pores, 10–16. Five subspecies in U.S., differing in adult males (females frequently indistinguishable subspecifically): (1) *S. m. magister*—broad, dark vertebral stripe, 4–5 scales wide, no bars or spots; (2) *S. m. bimaculosus*—2 rows of 6–7 paired dark spots on each side; (3) *S. m. transversus*—6–7 dark crossbands; (4) *S. m. uniformis*—light yellow or tan above; (5) *S. m. cephaloflavus*—head yellowish orange, back with 5–6 wavy crossbars.

Undulate Group: This group of spiny lizards consists of six small- to medium-sized species, all oviparous, with no distinctive characters. Wavy dark crossbands and a dorsolateral light stripe are characteristic, but vary. There are five species in the United States.

SAGEBRUSH LIZARD (*Sceloporus graciosus*). Small dorsal scales, keeled and pointed, 42–63 from head to above anus; 9–16 femoral pores. Scales on rear of thigh are granular. Three subspecies in U.S.: (1) *S. g. graciosus*—wavy dorsal bars well defined, males with blue areas on belly not fused with blue on throat; (2) *S. g. vandenburghianus*—like (1) but males more extensively blue below; (3) *S. g. arenicolus*—dorsal pattern lost except for light lines, throat not blue.

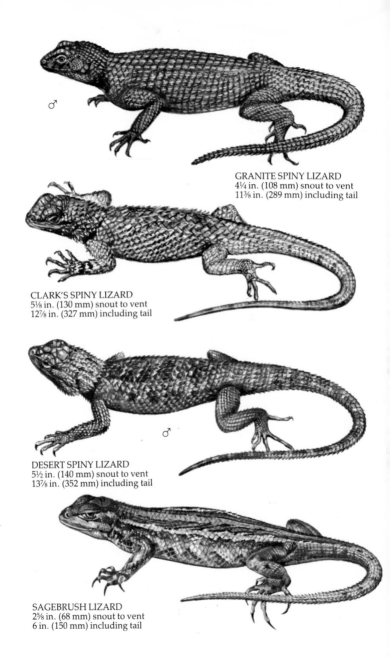

GRANITE SPINY LIZARD
4¼ in. (108 mm) snout to vent
11⅜ in. (289 mm) including tail

CLARK'S SPINY LIZARD
5⅛ in. (130 mm) snout to vent
12⅞ in. (327 mm) including tail

DESERT SPINY LIZARD
5½ in. (140 mm) snout to vent
13⅞ in. (352 mm) including tail

SAGEBRUSH LIZARD
2⅝ in. (68 mm) snout to vent
6 in. (150 mm) including tail

121

STRIPED PLATEAU LIZARD (*Sceloporus virgatus*). Has pale blue spot at rear sides of throat, not dark-bordered, surrounded or replaced by orange in breeding female. No blue at sides of belly in male.

FLORIDA SCRUB LIZARD (*Sceloporus woodi*). Males have pale blue area at sides of belly, narrowly black-bordered; small pale blue spot at each side of throat, surrounded by black. Females have some blue on sides of belly and throat, no black.

EASTERN FENCE LIZARD (*Sceloporus undulatus*). Highly variable pattern of wavy lines on brown or gray background. In males, sides of belly are blue, dark-bordered; there are 2 blue dark-bordered areas (sometimes fused) at rear of throat in all except 1 subspecies. Eight subspecies in U.S: (1) *S. u. undulatus*—37 or fewer dorsal scales from head to above anus (more in others); (2) *S. u. cowlesi*—pale, essentially patternless; (3) *S. u. garmani*—stripes distinct, no blue on throat; (4) *S. u. elongatus*—45 or more dorsal scales (fewer in others); (5) *S. u. tristichus*—belly patches in both sexes; (6) *S. u. erythrocheilus*—lips and chin yellow to red, especially in breeding season; (7) *S. u. hyacinthinus*—paravertebral marks expanded as many lines crossing back, when visible, and dorsolateral light lines indistinct in males; (8) *S. u. consobrinus*—paravertebral marks are small spots, dorsolateral lines distinct.

WESTERN FENCE LIZARD (*Sceloporus occidentalis*). Highly variable in appearance but basically has a series of paravertebral dark spots, broad and wavy or narrow and rounded, a vague dorsolateral light line, and a series of lateral spots, fused or indistinct. In adult males, pattern is more obscure, with numerous light flecks of blue or green. Six subspecies: (1) *S. o. occidentalis*—paired blue gular and belly patches large, ventrals (postmental to anus) 85 or fewer; (2) *S. o. bocourti*—like (1) except gular patches small in males (20 or fewer scales at least half blue); (3) *S. o. becki*—chin, lower lips barred (not in others); (4) *S. o. taylori*—ventral surfaces blue; (5) *S. o. longipes*—dorsals 45 or fewer, ventrals 85 or fewer, 1 gular patch; (6) *S. o. biseriatus*—like (5) but 46 or more dorsals, 86 or more ventrals.

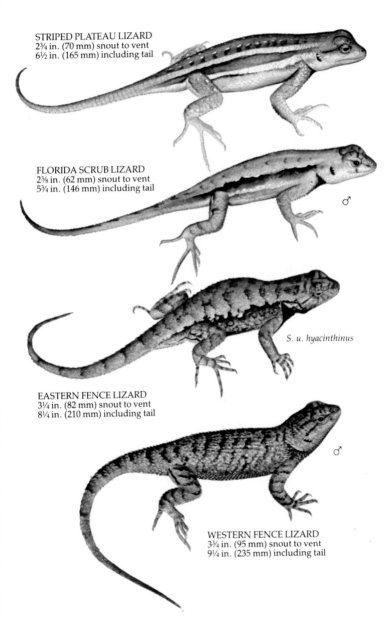

STRIPED PLATEAU LIZARD
2¾ in. (70 mm) snout to vent
6½ in. (165 mm) including tail

FLORIDA SCRUB LIZARD
2⅜ in. (62 mm) snout to vent
5¾ in. (146 mm) including tail

♂

S. u. hyacinthinus

EASTERN FENCE LIZARD
3¼ in. (82 mm) snout to vent
8¼ in. (210 mm) including tail

♂

WESTERN FENCE LIZARD
3¾ in. (95 mm) snout to vent
9¼ in. (235 mm) including tail

HORNED LIZARDS—genus *Phrynosoma*

Conspicuous features are a flattened, broad body, a short tail, and horns on the head. Ear openings and tympanums are visible in some, covered by granular skin in others. The femoral pores are best developed in males, with postanals enlarged in some. Horned lizards are primarily but not exclusively ant eaters. Their defensive behavior involves flattening the body, stiffening the legs, opening the mouth and hissing, rocking, jumping toward the threat, and squirting blood from the eyes (to 2–3 feet; 600–900 mm). They rely mostly on their concealing coloration. All U.S. species are oviparous except the Short-horned, which lives at high altitudes. About 13 species are known, 7 in the U.S.

KEY TO HORNED LIZARDS

1. 4 large occipital horns . . . .
. . . . **Regal,** below
   2 horns, large or small . . . . . *see* **2**
2. 3 or more paramedian rows of enlarged scales on throat . . . .
. . . . **Coast,** below
   2 or fewer rows . . . . *see* **3**
3. No fringe of enlarged scales at sides . . . . . . **Roundtail,** p. 126
   1 or 2 fringes . . . . *see* **4**

4. Rear lower labials much larger than adjacent chin shields . . . .
. . . . **Short-horned,** p. 126
   Converse . . . . *see* **5**
5. Dark vertebral streak . . . .
. . . .**Flat-tail,** p. 126
   Light or no streak . . . . *see* **6**
6. 1 peripheral fringe .**Desert,** p. 126
   2 peripheral fringes .**Texas,** below

TEXAS HORNED LIZARD (*Phrynosoma cornutum*). Two large occipital horns, not in series or parallel with temporal horns; ventrals keeled; 1 row of enlarged gulars on each side of throat between chin shield series.

COAST HORNED LIZARD (*Phrynosoma coronatum*). Two large parallel occipital horns, somewhat out of line with temporal horns. Rear temporals are nearly as large as occipitals. Has 2 peripheral fringes of spines; ventrals smooth. Large dark blotch on either side of nape; 3 pairs of wavy dark spots on back, light-edged to rear. Mottled below. Two subspecies in U.S.: (1) *P. c. blainvillei*—central interorbital scales enlarged, smooth; (2) *P. c. frontale*—scales not enlarged, rough.

REGAL HORNED LIZARD (*Phyrnosoma solare*). Four large occipital horns in contact with each other, in series with temporal horns; 1–2 series of enlarged gulars on each side. Rear chin shields are larger than lower labials. There is a single lateral fringe. Large middorsal oval light area, with 3 irregular crossbands, usually a light vertebral stripe; sides of back are darker; tail is banded. White below, with small dark spots.

124

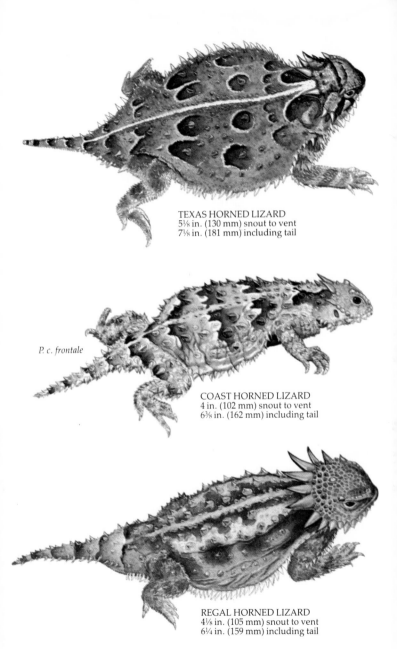

TEXAS HORNED LIZARD
5⅛ in. (130 mm) snout to vent
7⅛ in. (181 mm) including tail

P. c. frontale

COAST HORNED LIZARD
4 in. (102 mm) snout to vent
6⅜ in. (162 mm) including tail

REGAL HORNED LIZARD
4⅛ in. (105 mm) snout to vent
6¼ in. (159 mm) including tail

125

SHORT-HORNED LIZARD (*Phrynosoma douglassi*). Two small occipital horns, no larger than rear temporal horns; no gulars enlarged. Rear chin shields are smaller than rear scales behind lower labials. There is a single peripheral fringe of spines; ventrals are smooth. Large dark blotch on sides of neck, 3 pairs on body, light-bordered; head is reddish. Five subspecies in U.S.: (1) *P. d. douglassi*—small, dark-colored but clearly patterned, horns as long as broad, occipitals vertical; (2) *P. d. ornatum*—like (1) but pale, pattern weak; (3) *P. d. brevirostre*—horns shorter than broad, dark spots not light-edged or only at rear; (4) *P. d. ornatissimum*—horns shorter than broad, dark spots light-edged medially and posteriorly; (5) *P. d. hernandezi*—horns longer than broad.

FLAT-TAIL HORNED LIZARD (*Phrynosoma mcalli*). Has very long occipital horns, essentially in series and parallel with temporals; 1 row of slightly enlarged gulars on each side of throat. Rear chin shields are larger than lower labials; tympanums and ear openings not visible; ventrals feebly keeled. There are 2 peripheral fringes, the lower quite small.

ROUNDTAIL HORNED LIZARD (*Phrynosoma modestum*). Has occipital horns no more than half again as long as wide, little larger than rear temporals; no gulars enlarged. Chin shields contact entire length of lower labials (unique), rear ones much larger than lower labials. There are no visible ear openings or tympanums, and no lateral fringe.

DESERT HORNED LIZARD (*Phrynosoma platyrhinos*). Moderately large occipital horns, parallel with temporal horns; 1 row of slightly enlarged gulars on each side. Rear chin shields are much larger than rear lower labials. Ear openings can be concealed or not. Has single lateral fringe and smooth ventrals. Ground color much like that of local habitat, but has conspicuous lateral blotches on neck and 3–4 undulate paravertebral spots, light-edged to rear. Tail is banded. White below, with small scattered black spots. Two subspecies: (1) *P. p. platyrhinos*—occipital horns less than 45% of head length, space between same as basal width of 1 horn; (2) *P. p. calidiarum*—horns more than 45% of head length, space between half of basal width of 1 horn.

126

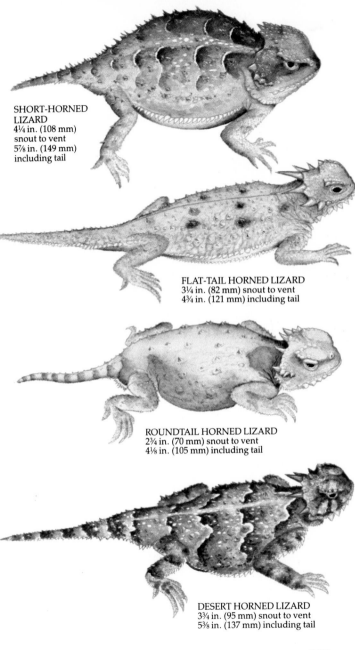

SHORT-HORNED
LIZARD
4¼ in. (108 mm)
snout to vent
5⅞ in. (149 mm)
including tail

FLAT-TAIL HORNED LIZARD
3¼ in. (82 mm) snout to vent
4¾ in. (121 mm) including tail

ROUNDTAIL HORNED LIZARD
2¾ in. (70 mm) snout to vent
4⅛ in. (105 mm) including tail

DESERT HORNED LIZARD
3¾ in. (95 mm) snout to vent
5⅜ in. (137 mm) including tail

127

SAND LIZARDS

These closely related genera share adaptations to life on sand. Oblique overlapping labials, granular or minute dorsal scales, smooth flat ventrals, small head scales, a granular gular fold, triangular median postmental, prominent postlabials, a more or less countersunk lower jaw, a nasal sand trap, long hind limbs, and flattened body and tail. Males have femoral pores and enlarged postanals. Oviparous.

EARLESS LIZARDS—genus *Holbrookia*

All three known species occur in the United States. No ear openings.

KEY TO EARLESS LIZARDS

1. Several black spots under tail . . . .
. . . . **Spot-tailed,** p. 130
No black spots . . . . *see* **2**

2. Dorsal scales distinctly keeled . . .
. . . . **Keeled,** below
Scales smooth . . . . **Lesser,** below

LESSER EARLESS LIZARD (*Holbrookia maculata*). Has very small, flat, smooth dorsal and lateral scales; ventrals larger, smooth. Femoral pores, 7–16. Pattern and color are variable; basically light middorsally, with dorsolateral and lateral light stripes, usually dim and often almost totally obscured. Has 8–14 paravertebral crescentic dark spots with indistinct anterior and more distinct, light-spotted posterior borders; similar series between dorsolateral and lateral light lines, with 3–5 on base of tail. Sides have 2 (rarely 3–4) oblique black bars surrounded by blue patch in males and often reduced in females. No marks under tail. Spotted-striped pattern replaced with light dots in males.

Six subspecies in U.S.: (1) *H. m. maculata*—tail shorter than snout-to-vent length (also in 2, 3, 4), stripes pale, male speckled; (2) *H. m. perspicua*—stripes pale, male not speckled; (3) *H. m. approximans*—stripes scarcely evident, male strongly speckled; (4) *H. m. ruthveni*—color strongly faded, whitish; (5) *H. m. pulchra*—larger, tail longer than snout-to-vent length (also in 6), pattern moderately distinct, male less densely speckled; (6) *H. m. thermophila*—stripes scarcely or not evident, male densely speckled.

KEELED EARLESS LIZARD (*Holbrookia propinqua*). Tail is as long as or longer than head-to-body length (not in Lesser, except subspecies in Southwest). Dorsal scales are small, keeled, pointed; laterals similar but smaller; ventrals smooth, flat, 3–4 times larger than dorsals. One subspecies in U.S.: *H. p. propinqua*.

H. m. maculata

LESSER EARLESS LIZARD
3 in. (76 mm) snout to vent
7½ in. (191 mm) including tail

H. m. thermophila

KEELED EARLESS LIZARD
2⅜ in. (62 mm) snout to vent
5½ in. (140 mm) including tail

SPOT-TAILED EARLESS LIZARD (*Holbrookia lacerata*). Dark spots under tail (unique in genus). Tail rarely longer than head and body. Dorsal and lateral scales small, flat, smooth; weakly keeled on parts of body in young; ventrals larger, smooth. Femoral pores, 21–41. Two subspecies: (1) *H. l. lacerata*—dark spots mostly fused as bands; (2) *H. l. subcaudalis*—spots mostly separate.

GREATER EARLESS—genus *Cophosaurus*

GREATER EARLESS LIZARD (*Cophosaurus texanus*). Resembles Zebratail but no ear opening. Dorsal scales minute; ventrals larger, all smooth. Black bars under tail. Two subspecies in U.S.: (1) *C. t. texanus*—27 or fewer femoral pores; (2) *C. t. scitulus*—28 or more.

ZEBRATAIL LIZARDS—genus *Callisaurus*

ZEBRATAIL LIZARD (*Callisaurus draconoides*). Has granular, smooth dorsal and lateral scales; large ear opening. Tail is crossbanded above and below. Three subspecies in U.S.: (1) *C. d. myurus*—tail 57% or less of total length, hind leg 91% or less of body length, 16 or fewer femoral pores, interparietal and supraorbital semicircles completely separated; (2) *C. d. rhodostictus*—58% or more, 92% or more, 16 or fewer, completely separated; (3) *C. d. ventralis*—56% or less, 91% or more, 17 or more, not completely separated.

FRINGE-TOED LIZARDS—genus *Uma*

Three species in the United States; two in Mexico. Small interparietal, extensive fringes on third and fourth toes, crescent-shaped nostrils, enlarged scales on rear of thigh, and long scales protecting ear openings. Dorsals minute, ventrals larger—both smooth. Oviparous.

KEY TO FRINGE-TOED LIZARDS

1. 1 rounded black spot on each side of belly . . . . *see* **2**
No black spot . . **Coachella,** p. 132

2. Convergent lines on throat expanded medially, united to form crescents . . . . . **Mojave,** below
Lines disappearing medially . . . .
. . . . **Fringe-toed,** p. 132

MOJAVE FRINGE-TOED LIZARD (*Uma scoparia*). Has 5 rows of scales between nostrils; 22–47 femoral pores; fringe spines on 4th toe, usually 33 or more. In breeding season, yellow-green below, pink on sides.

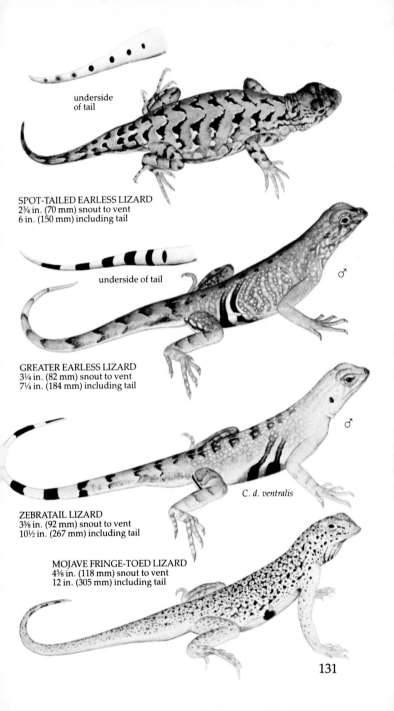

underside
of tail

SPOT-TAILED EARLESS LIZARD
2¾ in. (70 mm) snout to vent
6 in. (150 mm) including tail

underside of tail

♂

GREATER EARLESS LIZARD
3¼ in. (82 mm) snout to vent
7¼ in. (184 mm) including tail

♂

ZEBRATAIL LIZARD
3⅝ in. (92 mm) snout to vent
10½ in. (267 mm) including tail

C. d. ventralis

MOJAVE FRINGE-TOED LIZARD
4⅝ in. (118 mm) snout to vent
12 in. (305 mm) including tail

131

COACHELLA FRINGE-TOED LIZARD (*Uma inornata*). Has 3 rows of internasals; 18–28 femoral pores; fringe spines on the 4th toe, usually 32 or fewer. In breeding season, pinkish on sides of belly, with lips and rear of eyelids orange.

FRINGE-TOED LIZARD (*Uma notata*). Like Coachella but has conspicuous black spot at sides of midbelly. Two subspecies: (1) *U. n. notata*—belly spots in orange patch; (2) *U. n. rufopunctata*—patch around belly spots is pink in breeding season only.

TROPIDURINE LIZARDS—subfamily Tropidurinae

Many genera occur in South America, the Galapagos Islands, and the West Indies, where they are as diverse as sceloporines in North America. These lizards have a large interparietal, no femoral pores, and enlarged scales in the middorsal row.

CURLTAIL LIZARD (*Leiocephalus carinatus*). Rather large, keeled, pointed dorsal scales, 51–55 from head to above anus. Tail is slightly compressed, with tip habitually curled upward. One subspecies in U.S.: *L. c. armouri*—from the Bahamas.

AMPHISBAENIDS—suborder Amphisbaenia

This suborder contains 23 genera and about 140 species, all tropical or subtropical. All are soft-bodied, with numerous body rings bearing smooth, flat, rectangular scales. The tail is short and blunt; the eyes are concealed. There are no visible ear openings or, in most, legs. Some have preanal pores. Most are viviparous. This suborder contains three families, one with two subfamilies. The one species in our area belongs to the subfamily Rhineurinae of the family Amphisbaenidae.

FLORIDA WORM LIZARD (*Rhineura floridana*). No visible eyes. Lower jaw is countersunk; head is sharp-edged at jaw level, rounded in dorsal profile and in lateral profile above jaw level. There are no preanal pores. Dorsal surface of the tail is flattened and covered with tubercular scales. About 250 rings on body, each with 24–36 scales. Oviparous. Rose-colored in life, turns yellow-brown to white in preservatives.

COACHELLA FRINGE-TOED LIZARD
4⅝ in. (118 mm) snout to vent
12 in. (305 mm) including tail

FRINGE-TOED LIZARD
4⅝ in. (118 mm) snout to vent
12 in. (305 mm) including tail

CURLTAIL LIZARD
4⅛ in. (105 mm) snout to vent
10¼ in. (260 mm) including tail

FLORIDA WORM LIZARD
16 in. (406 mm) total length

133

SNAKES—suborder Serpentes

Snakes are the most successful of all living reptiles. Though the total number of lizard species (3,012) exceeds snake species (2,125), lizards are more limited in distribution so that in most areas more kinds of snakes occur. In the United States, snakes and lizards can be contrasted as follows: native genera—48:28; total genera (including established introduced ones)—50:42; native species—115:91; total species—117:113. Snakes range farther north and to higher altitudes, and are adapted to a much wider range of conditions. No snakes are herbivorous, but they are more specialized for capturing particular kinds of prey. Many have modified teeth for injecting a poison; others are specialized for eating snails, slugs, earthworms, toads, or even other snakes. Most are oviparous, but some are viviparous. Like other members of the order, they have two copulatory organs, called hemipenes.

Fourteen families of snakes are commonly recognized; only five occur in the U.S. and are listed here in approximate phylogenetic order.

1. Bigjawed blind snakes (Leptotyphlopidae) are native to the southwestern United States (three species) and tropical and subtropical regions of both hemispheres.

2. Boas and pythons (Boidae), a largely tropical family, are represented in the western United States by two small species. Possibly other species have been introduced to southern Florida.

3. Advanced snakes (Colubridae) account for 75 percent of North American species and 78 percent of all snakes. A few species are deadly poisonous to people (none in U.S.)

4. Elapids (Elapidae) include snakes, sea snakes, cobras, and allies, all with fixed front fangs and very poisonous. Only coral snakes (two species) occur in the area covered by this book.

5. Vipers (Viperidae) include "true" (or pitless) vipers, restricted to the Eastern Hemisphere; and pit vipers, most abundant in the Western Hemisphere. A total of 17 species of rattlesnakes and moccasins (copperheads and cottonmouths), all very poisonous, occur in the area covered by this book.

In blind snakes, the eye is a minute black dot under the large scales of the head. All other snakes treated in this book have a transparent scale (brille) over the eye. Blind snakes have no enlarged ventrals. Boas have ventrals of intermediate size. All others have greatly enlarged ventrals.

KEY TO FAMILIES

1. Red, yellow, and black rings about body, red bordered by yellow . . . . **elapids**, p. 196
If red, yellow, and black rings, red bordered by black . . . . . . *see* **2**

2. Scales smooth, cycloid, of uniform size . . . . . **blind snakes**, p. 136
Not as above . . . . *see* **3**

3. Deep pit between nostril and eye . . . . **pit vipers**, p. 198
Not as above . . . . *see* **4**

4. Large scales on top of head . . . . **colubrids**, p. 140
Small scales on top of head . . . . **true boas**, p. 138

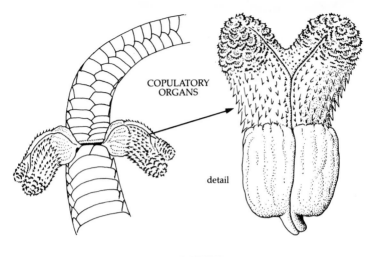

COPULATORY
ORGANS

detail

KEY TO FAMILIES

1

coral snake pattern: red, yellow, black rings—the red bordered by yellow

2

scales smooth, cycloid

3

loreal pit

large
scales

small
scales

4

colubrid

boa

135

BLIND SNAKES—family Leptotyphlopidae

Blind snakes, the most primitive of the snakes, are nocturnal burrowers. Leptotyphlopids are confined largely to American and African tropics and subtropics. The left oviduct is missing. Scales are uniform in size and shiny; ventrals are not enlarged. A spine is present on the tip of the tail. Coloration is a uniform purplish, brown, or gray; belly is often whitish. There are teeth in the lower jaw only; the skull bones are solidly joined.

Blind snakes feed on termites and ants and their larvae. All are oviparous except possibly one species of southern Asia. This family has two genera and about 60 species. All of the blind snakes treated in this book live in arid to semiarid regions or in sandy or gravelly soils of dry grasslands.

KEY TO BLIND SNAKES

1. 1 scale on top of head between oculars . . . . . . **Western,** below
3 scales . . . . *see* **2**

2. 2 upper labials between ocular and nasal . . . . **Texas,** below
1 upper labial . . **Mountain,** below

WESTERN BLIND SNAKE (*Leptotyphlops humilis*). Has 257–283 scales from head to tip of tail, 15–21 caudals. Sometimes has a purplish cast. Four subspecies in U.S.: (1) *L. h. humilis*—7 or more dorsal scale rows, heavily pigmented; (2) *L. h. cahuilae*—5 lightly pigmented dorsal scale rows; (3) *L. h. segregus*—10 scale rows around tail (12 in others); (4) *L. h. utahensis*—4th median scale from rostral usually divided, 5th much wider than 6th (neither true in others).

TEXAS BLIND SNAKE (*Leptotyphlops dulcis*). Has 206–255 scales from head to tip of tail, 12–17 caudals, 10 scale rows around tail. Range overlaps Mountain Blind Snake in some areas; interrelationship not clear.

MOUNTAIN BLIND SNAKE (*Leptotyphlops myopicus*). Has 244–246 scales from snout to tip of tail, 12–16 caudals, 10 scale rows around tail. Occurs mainly in high plains and in canyons on both sides of S Continental Divide. Where range overlaps Texas Blind Snake, upper labial character is usually reliable for differentiating species. Isolating mechanism is presumably a pheromone, but not established. One subspecies in U.S.: *L. m. dissectus*.

KEY TO BLIND SNAKES

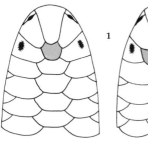

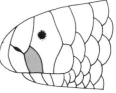

1 scale   3 scales

2 upper labials

1 upper labial

WESTERN BLIND SNAKE
16 in. (406 mm) total

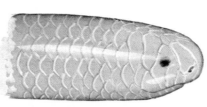

TEXAS BLIND SNAKE
10¾ in. (273 mm) total

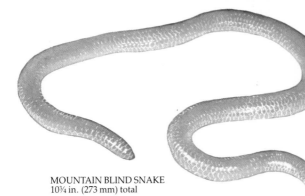

MOUNTAIN BLIND SNAKE
10¾ in. (273 mm) total

TRUE BOAS—family Boidae, subfamily Boinae

Two genera, each with a single species, are native to the United States. The Boa Constrictor of the American tropics occurs natively within 100 miles (160 kilometers) of the Mexican border on both east and west coasts and is frequently released in the southeastern United States. It is not known to reproduce successfully in the United States.

True boas are nocturnal and viviparous, varying from 2-foot (.6 m) burrowers to 38-foot (11.6 m) giants (reputedly). Some are slender-bodied and arboreal, others heavy-bodied and either terrestrial or aquatic. All are constrictors, subsisting primarily on warm-blooded prey. Most, if not all, possess remarkably sensitive heat-detecting sense organs along the lips, but pits housing the organs are evident in only a few. Eyes are small, with vertical pupils. Body scales are small, numerous, and smooth in North American species; ventrals are wider than dorsals. Minute clawlike vestiges of hind limbs, best developed in males, are visible on either side of anus.

KEY TO BOAS

1. Large scales on top of head, 3 between orbits . . . **Rubber,** below
   Not as above . . . . *see* **2**

2. 2 scales bordering rostral to rear . . . . **Rosy,** below
   Many scales . . . . **Boa Constrictor,** below

BOA CONSTRICTOR (*Boa constrictor*). Very small head scales; 65–95 scale rows; 225–261 ventrals, anal entire; 48–70 caudals, mostly entire. Eight subspecies, 2 imported freely: (1) *B. c. constrictor*—head line without projections toward eyes, ovoid dorsal spots not especially reddish, Amazonian S.A. to Argentina; (2) *B. c. imperator*—projections present, tail spots reddish, seldom over 7 ft. (2.1 m), Mexico to NW S.A. [No range map.]

ROSY BOA (*Lichanura trivirgata*). Somewhat enlarged head scales, 2 median pairs following rostral, 5–7 between orbits; 35–43 scale rows; 218–244 ventrals, anal entire; 39–51 caudals, entire. Three subspecies: (1) *L. t. trivirgata*—dark brown stripes on drab background, belly with few marks; (2) *L. t. gracia*—stripes lighter, even-edged, no spots between stripes, belly profusely marked; (3) *L. t. roseofusca*—stripes irregular-edged and spotted between, belly profusely marked.

RUBBER BOA (*Charina bottae*). Large head scales; 39–47 scale rows; 184–220 ventrals, anal entire; caudals undivided. Blunt tail; uniformly olive, brown, or tan. Young, 2 to 8.

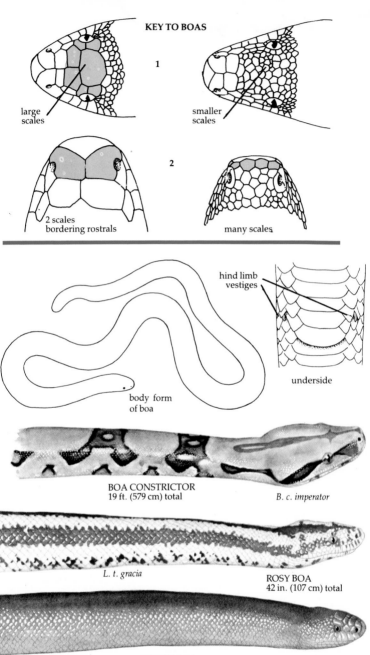

KEY TO BOAS

1

large scales

smaller scales

2

2 scales bordering rostrals

many scales

hind limb vestiges

body form of boa

underside

BOA CONSTRICTOR
19 ft. (579 cm) total

B. c. imperator

L. t. gracia

ROSY BOA
42 in. (107 cm) total

RUBBER BOA
33 in. (838 mm)

COLUBRID SNAKES—family Colubridae

This huge family (350 genera, 1,750 species) includes 92 U.S. species, most nonpoisonous. A few U.S. species have weak venom (little or no effect on humans) and fixed rear fangs; others are constrictors. Some are racers.

KEY TO COLUBRID SNAKES

1. 2 scales between eye, nostril . . . . *see* 2
   3 or more scales . . . . *see* 6
2. 13 scale rows . . . . . **Worm**, p. 162
   15–19 rows . . . . *see* 3
3. 15–17 scale rows . . . . *see* 4
   19 .**Mud, Rainbow snakes**, p. 162
4. Prefrontal contacts eye . . . . **earth snakes**, p. 152
   Not as above . . . . *see* 5
5. Scales keeled . . . . **brown snakes**, p. 158
   Smooth . . **flathead snakes**, p. 170
6. Some or all dorsals keeled (examine rear trunk) . . . . . . *see* 7
   All smooth . . . . *see* 17
7. Anal plate entire . . . . *see* 8
   Divided . . . . *see* 11
8. Prefrontals 4, 27 or more rows of scales . . . . . . **Bullsnake**, p. 186
   2, fewer than 27 . . . . *see* 9
9. Rostral huge; suboculars present . . . . . . **leafnose snakes**, p. 164
   Normal; no suboculars . . . .*see* 10
10. 8 or more lower labials . . . . **garter, ribbon snakes**, p. 144
    7 or fewer . . . . **Lined**, p. 152
11. Rostral keeled . . . . **hognose snakes**, p. 166
    Not keeled . . . . *see* 12
12. 1 internasal . . . . **Striped Swamp**, p. 158
    2 internasals . . . . *see* 13
13. 17 scale rows . . . . *see* 15
    More . . . . *see* 14
14. Keels to tip of all dorsals except sometimes lower row . . . .*see* 42
    Absent or reduced . . . . **rat snakes**, p. 184
15. 100 or more caudals . . . . . *see* 16
    Fewer than 50 . . . . **Black Swamp**, p. 158
16. 7 upper labials . . . . **green snakes**, p. 188
    9 . . . . **Speckled Racer**, p.188
17. Anal plate entire . . . . *see* 18

Divided . . . .*see* 25
18. All or most caudals divided . . . .*see* 19
    Not at above . . **Longnose**, p. 164
19. Eye pupil vertical . . . **Lyre**, p. 176
    Round . . . .*see* 20
20. Parietals touch upper labials . . . .**Short-tailed**, p. 178
    Not as above . . . .*see* 21
21. Belly unmarked . . . .*see* 22
    With dark marks, ...*see* 24
22. Rostral upturned . . . .**Desert Hooknose**, p. 174
    Not upturned . . . .*see* 23
23. 6–7 upper labials . . . . **Scarlet**, p. 178
    8 . . . . **Glossy**, p. 182
24. 17 rows of scales . . . . **Indigo**, p. 188
    More than 17 . **kingsnakes**, p. 178
25. 1 preocular . . . .*see* 26
    2 or more . . . .*see* 35
26. 13–17 rows of scales . . . . . .*see* 27
    19 or more . **Black-striped**, p. 176
27. 13–15 rows of scales . . . . . .*see* 31
    7 rows . . . .*see* 28
28. Rostral upturned . . . .*see* 29
    Not upturned . . . .*see* 30
29. Rostral contacts frontal . . . . **Mexican Hooknose**, p. 174
    Not as above . . . . **Western Hooknose**, p. 174
30. Prefrontal contacts upper labials . . . . **Mexican Vine**, p. 188
    Not as above .**Pine Woods**, p. 176
31. Internasals touch . . . . *see* 32
    Separated medially . . . .**Banded Sand**, p. 170
32. Black bar on each ventral . . . . **Sharptail**, p. 162
    Not as above . . . .*see* 33
33. Snout flat . . . .**shovelnose snakes**, p. 168
    Normal . . . .*see* 34
34. 65 or more caudals . . . . **green snakes**, p. 188
    59 or fewer .**ground snakes**, p. 166

140

.35. 19 or more scale rows . . . . . *see* **36**
     Fewer than 17 . . . . *see* **38**
36. 3–4 preoculars . . . . . **Lyre,** p. 176
     2 preoculars . . . . *see* **37**
37. Lateral spots . . . . **Night,** p. 176
     None . . . . **Cat-eyed,** p. 176
38. 1 anterior temporal . . . . . . *see* **39**
     2 or 3 . . . . *see* **40**
39. Light neck ring . . . .
     . . . . **Ringneck,** p. 160
     No ring . . . . **green snakes,** p. 188

40. Rostral large, free edges . . . .
     . . . . **patchnose snakes,** p. 194
     Normal . . . . *see* **41**
41. 15 rear body scale rows . . . .
     . . . . **Racer,** p. 190
     13 or fewer . . **whipsnakes,** p. 190
42. 19 rows of scales . . . . *see* **43**
     More . . . . . **water snakes,** p. 154
43. 7 or more upper labials . . . .
     . . . . . . **crayfish snakes,** p. 156
     6 or fewer . . . . **Kirtland's,** p. 156

KEY TO COLUBRID SNAKES

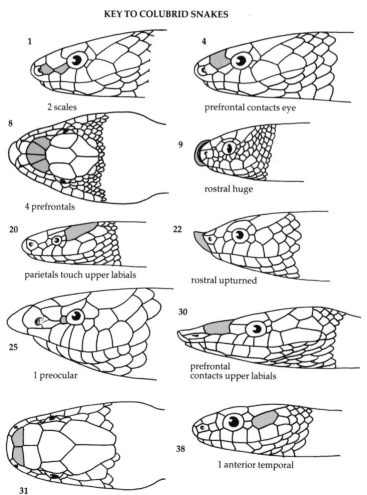

1
2 scales

4
prefrontal contacts eye

8
4 prefrontals

9
rostral huge

20
parietals touch upper labials

22
rostral upturned

25
1 preocular

30
prefrontal
contacts upper labials

31
internasals touch

38
1 anterior temporal

141

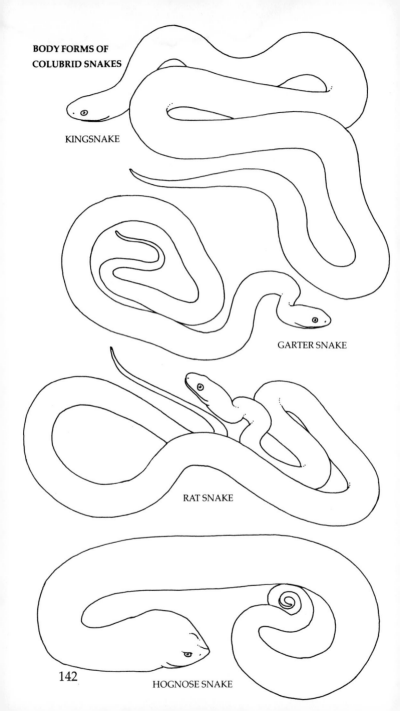

**BODY FORMS OF
COLUBRID SNAKES**

KINGSNAKE

GARTER SNAKE

RAT SNAKE

142

HOGNOSE SNAKE

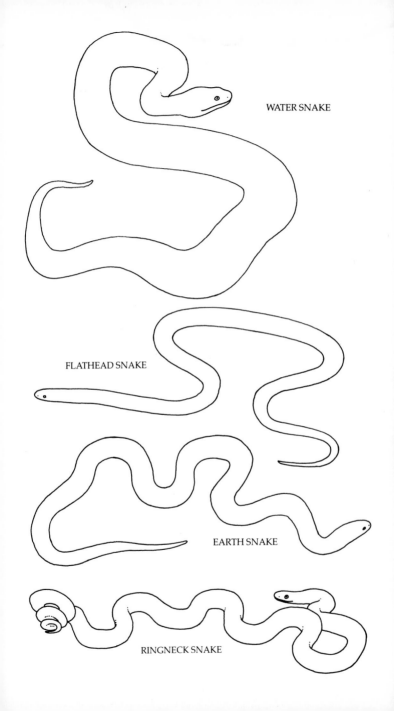

WATER SNAKE

FLATHEAD SNAKE

EARTH SNAKE

RINGNECK SNAKE

NATRICINE SNAKES—subfamily Natricinae

Internal characters define this subfamily. All are viviparous (all other U.S. colubrids are oviparous). Considered the most successful of all snakes, natricines are most abundant in number of individuals. None are poisonous or have enlarged teeth. Diet is mainly amphibians, fish, and worms or other invertebrates; rarely birds or mammals.

GARTER AND RIBBON SNAKES—genus *Thamnophis*

The most common snakes, all with keeled scales, undivided anal plate, and 1 anterior temporal. Of 22 known species, 13 are in the U.S.

KEY TO GARTER SNAKES AND RIBBON SNAKES

1. Lateral light line onto 4th scale row of forebody . . . . *see* **2**
   Line lower or absent . . . . . . *see* **8**
2. (*not shown*) Tail 27% or more of total length . . . . *see* **3**
   Tail shorter . . . . *see* **4**
3. (*not shown*) Ventrolateral dark stripe; E of Mississippi . . . .
   . . . . . . **Eastern Ribbon,** below
   No dark stripe; W of Mississippi .
   . . . . . **Western Ribbon,** p. 146
4. (*not shown*) Lateral light line onto 2nd scale row . . . . *see* **5**
   On 3rd and 4th rows only . *see* **7**
5. (*not shown*) 19 rows of scales at midbody . . . . *see* **6**
   17 rows .**Shorthead Garter,** p. 146
6. E of Mississippi . .**Butler's,** p. 146
   W of Mississippi .**Common,** p. 148
7. 8–9 upper labials . . . .
   . . . . **Mexican,** p. 146
   7 . . . . **Plains,** p. 146
8. (*not shown*) 17 rows of scales at midbody . . . . *see* **9**
   19–21 rows . . . . *see* **10**

9. (*not shown*) 7 upper labials . . . . . .
   . . **Northwestern Garter,** p. 150
   6 . . . . . .**Shorthead Garter,** p. 146
10. (*not shown*) Lateral light line on 3rd scale row of forebody . . . .
    . . . . **Checkered Garter,** p. 150
    On 2nd and 3rd rows, or absent . .
    . . . .*see* **11**
11. 7 upper labials . **Common,** p. 148
    8 upper labials . . . .*see* **12**
12. (*not shown*) Black blotch behind head on each side . . . .
    . . . . .**Blackneck Garter,** p. 150
    Not as above . . . .*see* **13**
13. 1 upper labial touching eye . . . . .
    . . . **Narrowhead Garter,** p. 152
    2 . . . .*see* **14**
14. Internasals broader than long, not tapered anteriorly; 6th and 7th upper labials enlarged . . . .
    . . . **Western Terrestrial,** p. 148
    Internasals longer than broad, tapered; labials not enlarged . . . .
    . . . . . **Western Aquatic,** p. 150

EASTERN RIBBON SNAKE (*Thamnophis sauritus*). Has 19 scale rows at midbody; 143–177 ventrals; 94–136 caudals; usually 7 (often 8) upper labials, 2 touching eye; 1 preocular; usually 3 postoculars. Lateral stripes on rows 3 and 4. Paired parietal spots absent or faint, not touching. Four U.S. subspecies: (1) *T. s. sauritus*—7 upper labials, stripes sharply defined, tail one-third total length; (2) *T. s. septentrionalis*—7 upper labials, median stripe obscure, tail less than one-third length; (3) *T. s. sackeni*—8 upper labials, obscure or no median stripe, tail over one-third length; (4) *T. s. nitae*—like (3) but has blue lateral stripe.

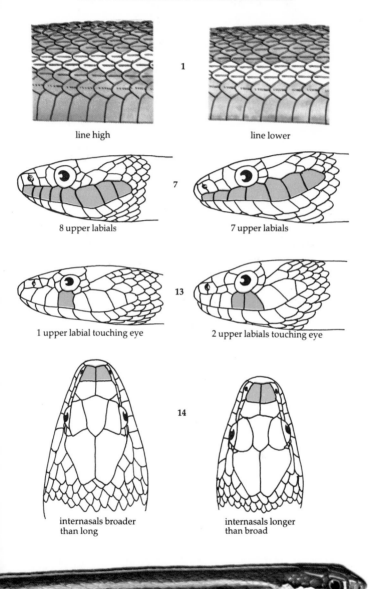

1

line high

line lower

7

8 upper labials

7 upper labials

13

1 upper labial touching eye

2 upper labials touching eye

14

internasals broader
than long

internasals longer
than broad

EASTERN RIBBON SNAKE
40 in. (102 cm) total

WESTERN RIBBON SNAKE (*Thamnophis proximus*). Has 19 rows of scales at midbody; 141–181 ventrals; 82–131 caudals; usually 7 (occasionally 8) upper labials light colored and unmarked, 2 contacting eye. Tail is 27% or more of total length. Much like Eastern Ribbon, but E of 98th meridian and has no ventrolateral dark stripe. Paired parietal spots are large, bright, touching. Four subspecies in U.S.: (1) *T. p. proximus* — back black, median stripe narrow, orange; (2) *T. p. orarius*—back olive-brown, stripe broad, gold; (3) *T. p. diabolicus*—back olive-gray to olive-brown, stripe orange, narrow ventrolateral dark stripe; (4) *T. p. rubrilineatus*—like (3) but stripe usually red, ventrolateral dark stripe often absent.

MEXICAN GARTER SNAKE (*Thamnophis eques*). Has 21–23 rows of scales at midbody; 149–172 ventrals; 68–69 caudals; 8–9 upper labials, 2 touching eye; 1 preocular; 3–4 postoculars. Lateral stripes on rows 3–4; concealed lateral black edges on ventrals. One subspecies in U.S.: *T. e. megalops*.

PLAINS GARTER SNAKE (*Thamnophis radix*). Has 19–21 (usually 19) rows of scales at midbody; 139–176 ventrals; 57–88 caudals; usually 7 (sometimes 8) upper labials, 2 touching eye; 1 preocular; 3 postoculars. Lateral stripes are on rows 3 and 4; lips are barred. Small dark spots found between and below stripes and on sides of belly. Two subspecies in U.S.: (1) *T. r. radix*—spots large, 154 or fewer ventrals, 19 rows of scales on rear neck; (2) *T. r. haydeni*—spots smaller, 155 or more ventrals, 21 rows of scales on rear neck.

BUTLER'S GARTER SNAKE (*Thamnophis butleri*). Has 19 rows of scales at neck and midbody, 129-147 ventrals, 51-71 caudals. Head is small, not wider than neck; usually has 7 (often 6) upper labials, 2 touching eye; 1 preocular; usually 3 (often 2) postoculars. Lateral stripes are present on scale rows 2, 3, and 4. Spots are obscure or indiscernible.

SHORTHEAD GARTER SNAKE (*Thamnophis brachystoma*). Has 17 rows of scales at neck and midbody, 132–146 ventrals, 51–72 caudals. Head is small, not wider than neck; usually 6 upper labials. Lateral stripes are on rows 2 and 3 (sometimes on 4), often with black border. No spots are evident.

146

WESTERN RIBBON SNAKE
48 in. (122 cm) total

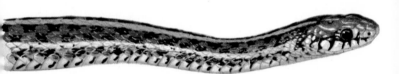

MEXICAN GARTER SNAKE
40 in. (102 cm) total

PLAINS GARTER SNAKE
41 in. (104 cm) total

BUTLER'S GARTER SNAKE
27 in. (686 mm) total

SHORTHEAD GARTER SNAKE
22 in. (559 mm) total

147

COMMON GARTER SNAKE (*Thamnophis sirtalis*). This is the most widely distributed snake in the U.S. The Common Garter Snake has 19 rows of scales at midbody, 137–178 ventrals, 54–97 caudals, 7 upper labials, 1 preocular, and 3 postoculars. Lateral stripes are present on rows 2 and 3 (also on row 4 in 1 subspecies). There are twelve subspecies in the U.S.

EASTERN: (1) *T. s. sirtalis*—no red between stripes; (2) *T. s. parietalis*—lips barred, red between broad stripes, belly marked with black; (3) *T. s. annectans*—no red, broad orange middorsal stripe; (4) *T. s. similis*—dark between stripes, spots obscured, lateral stripes bluish, dorsal tan or yellow; (5) *T. s. semifasciatus*—like *T. s. sirtalis* but anterior spots are fused into blotches; (6) *T. s. pallidula*—like *T. s. sirtalis* but with prominent spotting, median stripe absent or faint.

WESTERN: (1) *T. s. dorsalis*—like *T. s. parietalis* but pattern dull; (2) *T. s. infernalis*—like *T. s. parietalis* but lips are flecked with black; (3) *T. s. tetrataenia*—spots between stripes are fused into 2 black stripes, red between; (4) *T. s. fitchi*—like *T. s. parietalis* but top of the head and tips of ventrals are black, median stripe is broad; (5) *T. s. pickeringi*—like *T. s. fitchi* but median stripe limited to 1 row; (6) *T. s. concinnus*—like *T. s. parietalis* but top of the head is red, areas between and below stripes black onto belly except for red bars.

WESTERN TERRESTRIAL GARTER SNAKE (*Thamnophis elegans*). This snake has 19–21 rows of scales at midbody, 146–185 ventrals, and 67–101 caudals. Internasals are broader than long, not pointed anteriorly. It usually has 8 (occasionally 7) upper labials, 6 and 7 enlarged and often higher than wide. Usually has 1–2 preoculars and 3–4 postoculars. Lateral stripes are on 2nd and 3rd scale rows. There are four subspecies in the U.S.: (1) *T. e. elegans*—median stripe broad, sharp-edged, lateral stripes fairly distinct, area between black, few light flecks, belly pale; (2) *T. e. biscutatus*—like (1) but belly is gray, suffused with black toward rear; (3) *T. e. terrestris*—irregular dark spotting between median and lateral stripes, red-orange flecks on belly and sides, median stripe as in (1); (4) *T. e. vagrans*—dorsal stripe narrower, not sharp-edged, spotting present but no red.

T. s. sirtalis

COMMON GARTER SNAKE
52 in. (132 cm) total

T. s. tetrataenia

T. s. concinnus

T. e. terrestris

WESTERN TERRESTRIAL GARTER SNAKE
42 in. (107 cm) total

T. e. vagrans

WESTERN AQUATIC GARTER SNAKE (*Thamnophis couchi*). Has 19–21 rows of scales at midbody, 140–184 ventrals, 61–97 caudals. Internasals longer than broad, pointed anteriorly. Has 8 upper labials, 6 and 7 not enlarged; 1 or 2 preoculars; 3 postoculars. Lateral stripes weak or absent, on rows 2 and 3. Six subspecies in U.S.: (1) *T. c. couchi*—stripes indiscernible, broken by irregular dark spotting on gray body, 11 lower labials; (2) *T. c. hydrophila*—like (1) but 10 lower labials; (3) *T. c. gigas*—stripes dull yellow with diffuse margins, small alternating spots confined to area between stripes; (4) *T. c. aquaticus*—median stripe broad, sharp-edged, yellow to orange, spots small, separate, irregular, throat yellow; (5) *T. c. atratus*—like (4) but spots larger, extensively fused, median stripe a little narrower; (6) *T. c. hammondi*—no dorsal stripe or confined to neck only, lower row of spots only, lateral stripes narrow and diffuse, range overlaps with (5) in central coastal California.

NORTHWESTERN GARTER SNAKE (*Thamnophis ordinoides*). Has 17 rows of scales at midbody, 137–162 ventrals, 54–81 caudals, usually 7 upper and 8–9 lower labials, 1 (occasionally 2) preoculars, 3 postoculars. Median stripe is broad, yellow to red, sometimes dim or absent; lateral stripes, on rows 2 and 3, are distinct or faint. Spots are dim, separate. Ground color is black, brown, red, greenish, or bluish; belly is yellowish to slate, often with red blotches, sometimes marked with black.

CHECKERED GARTER SNAKE (*Thamnophis marcianus*). Has 21 rows of scales at midbody, 139–173 ventrals, 61–83 caudals, 8 upper labials, 1 preocular, 3–4 postoculars. Dark spots in 2 rows, with large and invading stripes. Conspicuous light crescent and black spot behind mouth. One subspecies in U.S.: *T. m. marcianus*.

BLACKNECK GARTER SNAKE (*Thamnophis cyrtopsis*). Has 19 rows of scales at midbody, 148–183 ventrals, 63–102 caudals, 8 upper labials, 1 preocular, 3 postoculars. Stripes are broad, distinct; median stripe is orange anteriorly, yellowish to rear; the laterals, on 2nd and 3rd scale rows, yellowish to tan. Black blotch on each side behind head. Belly unmarked, whitish to greenish. Two subspecies in U.S.: (1) *T. c. cyrtopsis*—dark spots in 2 rows on each side, not fused; (2) *T. c. ocellatus*—dark spots fused as 1 row anteriorly.

T. c. hydrophilia

WESTERN AQUATIC GARTER SNAKE
57 in. (145 cm) total

T. c. aquaticus

NORTHWESTERN GARTER SNAKE
26 in. (660 mm) total

CHECKERED GARTER SNAKE
42½ in. (108 cm) total

BLACKNECK GARTER SNAKE
42 in. (107 cm) total

NARROWHEAD GARTER SNAKE (*Thamnophis rufipunctatus*). Has 21 rows of scales at midbody; 152–177 ventrals; 65–87 caudals; 8–9 upper labials, 1 touching eye; 2–3 preoculars; and 2–4 postoculars.

LINED SNAKES—genus *Tropidoclonion*

LINED SNAKE (*Tropidoclonion lineatum*). A diminutive derivative of garter snakes, with keeled scales, undivided anal plate, and single anterior temporal. Diagnostic are reduced number of upper labials (usually 5 or 6) and double row of half-moon-shaped, sharply defined black spots down middle of belly and tail. Usually 19 rows of scales at midbody, 132–156 ventrals, 24–47 caudals, 1 preocular, 2 postoculars. Lateral light line on 2nd and 3rd row of scales (sometimes only 2nd) anteriorly; dark area between stripes is without dark spots. Four subspecies in U.S.: (1) *T. l. lineatum*—143 or fewer ventrals, 33 or more caudals in females, 41 or more in males; (2) *T. l. annectens*—144 or more ventrals, 34 or more caudals in females, 41 or more in males; (3) *T. l. texanum*—143 or fewer ventrals, 33 or fewer caudals in females, 40 or fewer in males; (4) *T. l. mertensi*—142 or fewer ventrals, 15 or 16 scale rows at rear of body (17 in others except males of 2 and 3), 41 or more caudals in males, 30–37 in females.

EARTH SNAKES—genus *Virginia*

These diminutive burrowers have pointed, conical heads and little pattern. Rows of scales at midbody number 15–17, anal divided, no preocular, one anterior temporal. Smooth has six upper labials, usually two postoculars, two internasals, and smooth or only weakly keeled dorsals. Rough has five upper labials, usually one postocular, one internasal, and prominently keeled dorsals.

ROUGH EARTH SNAKE (*Virginia striatula*). Has 17 rows of scales at midbody. Young usually have a light ring or pair of spots at rear of the head, lost in adults.

SMOOTH EARTH SNAKE (*Virginia valeriae*). Sometimes has faint median light line. Three subspecies: (1) *V. v. valeriae*—15 rows of scales at midbody, smooth except near tail, tiny black spots on back and sides; (2) *V. v. pulchra*—17 rows of scales at midbody, 15 anteriorly, weakly keeled; (3) *V. v. elegans*—like (2) except 17 rows of scales anteriorly.

NARROWHEAD GARTER SNAKE
34 in. (864 mm) total

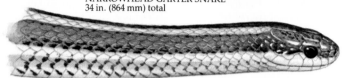

LINED SNAKE
21 in. (533 mm) total

underside

ROUGH EARTH SNAKE

1 internasal

1 postocular

SMOOTH EARTH SNAKE

2 internasals

2 postoculars

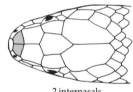

ROUGH EARTH SNAKE
12¾ in (324 mm) total

SMOOTH EARTH SNAKE
13¼ in. (337 mm) total

WATER SNAKES—genus *Nerodia*

This genus occurs in the U.S. and western Mexico. The 7 species in our area are alike in living in or near water, preference for vertebrate food (mostly fish and amphibians), intractability, odorous secretions when first captured, small eyes, tapered snout, ability to flatten head and body, heavily keeled scales in 21–33 rows, divided anal, usually 8 labials, 1 anterior temporal, 1 preocular, and 2–4 postoculars.

KEY TO WATER SNAKES

1. Eye completely separated from lip scales . . . . **Green,** below
   Only partially separated . . . *see* **2**
2. 2 rows of dark spots on each side, no median row . **Harter's,** p. 156
   Not as above . . . . *see* **3**
3. 2 anterior temporals . . . . **Brown,** below
   1 anterior temporal . . . . *see* **4**
4. Spots in median and lateral rows alternating, widely separated but interconnected at corners . . . . **Diamondback,** below

Not as above . . . . *see* **6**
5. Striped; or tail compressed at base; or dark streak from eye to corner of mouth . . . . **Southern,** p. 156
   Not as above . . . . *see* **6**
6. Belly not patterned, or discolored only along bases at edges of ventrals . . . . **Plainbelly,** below
   Boldly patterned . . . . **Northern,** p. 156

GREEN WATER SNAKE (*Nerodia cyclopion*). Has 27–31 rows of scales at midbody, 132–148 ventrals, 57–84 caudals, 1–2 preoculars and postoculars. Two subspecies: (1) *N. c. cyclopion*—belly mostly brown; (2) *N. c. floridana*—belly mostly white.

BROWN WATER SNAKE (*Nerodia taxispilota*). Has 27–33 rows of scales, 136–152 ventrals, 70–99 caudals, and 2-4 anterior temporals (usually 1 in others).

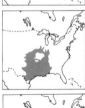

DIAMONDBACK WATER SNAKE (*Nerodia rhombifera*). Has 24–28 (usually 25–27) rows of scales, 135–146 ventrals, 59–82 caudals, usually 3 postoculars. Numerous tubercles on male's chin (unique). One subspecies in U.S.: *N. r. rhombifera*.

PLAINBELLY WATER SNAKE (*Nerodia erythrogaster*). Has 20–25 (usually 23) rows of scales, 145–156 ventrals, 59–84 caudals, usually 3 postoculars. Young have 29–35 large median spots, narrow lateral bars. Four subspecies: (1) *N. e. erythrogaster*—belly bright reddish, ground color scarcely reaching ends of ventrals; (2) *N. e. neglecta*—like (1) except darker, ground color onto ends of ventrals; (3) *N. e. flavigaster*—like (1) but belly yellow; (4) *N. e. transversa*—like (3) but dorsal pattern discernible in adults.

154

KEY TO WATER SNAKES

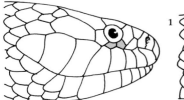

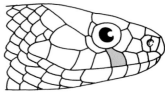

eye separated from lip scales

only partially separated

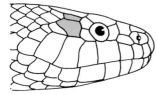

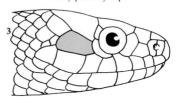

2 anterior temporals

1 anterior temporal

GREEN WATER SNAKE
74 in. (188 cm) total

N. c. floridana

BROWN WATER SNAKE
69 in. (175 cm) total

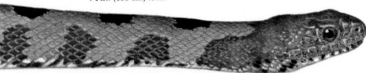

DIAMONDBACK WATER SNAKE
63 in. (160 cm) total

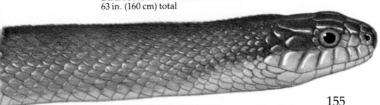

PLAINBELLY WATER SNAKE
62 in. (158 cm) total

N. e. flavigaster

NORTHERN WATER SNAKE (*Nerodia sipedon*). Has 23–25 rows of scales, 127–149 ventrals, 56–84 caudals, usually 3 postoculars. Four subspecies: (1) *N. s. sipedon*— lateral dark bars wider than interspaces, pattern distinct; (2) *N. s. pleuralis*—bars narrower and distinct; (3) *N. s. insularum*—pattern indistinct, light-colored; (4) *N. s. williamengelsi*—indistinct pattern above, dark.

SOUTHERN WATER SNAKE (*Nerodia fasciata*). Has 23 (occasionally 25) scale rows, 121–136 ventrals, 61–86 caudals, 3 (sometimes 2) postoculars. Six subspecies: (1) *N. f. fasciata*—19 or more crossbands, squarish marks on belly; (2) *N. f. confluens*—like (1) but 17 or fewer; (3) *N. f. pictiventris*—wavy crossbars on belly; (4) *N. f. clarki*—5 light and 4 dark stripes full length of body, central row of light spots on dark belly, base of tail compressed, living in brackish water or saltwater; (5) *N. f. taeniata*—like (4) but stripes short, blotched on rest of body; (6) *N. f. compressicauda*—like (4) but no central row of light belly spots, unicolor or blotched above, no stripes or on neck only.

HARTER'S WATER SNAKE (*Nerodia harteri*). Has 21 (sometimes 23) rows of scales, 143–151 ventrals, 64–88 caudals, 2–3 postoculars. Two subspecies: (1) *N. h. harteri*—2 rows of scales between posterior chin shields; (2) *N. h. paucimaculata*—1 row, spots on belly absent or reduced in size, number, and distinctness.

KIRTLAND'S SNAKES—genus *Clonophis*

KIRTLAND'S SNAKE (*Clonophis kirtlandi*). A specialized derivative of water snakes. Has 19 rows of scales (keeled) at midbody, 121–136 ventrals, 54–69 caudals, divided anal, 1 preocular, 2 postoculars, usually 6 (often 5) upper labials, 1 anterior temporal.

CRAYFISH SNAKES—genus *Regina*

These relatives of water snakes eat crayfish, have 19 rows of keeled scales, a divided anal, 7 upper labials, 1 anterior temporal, and usually 2 preoculars. The stripes fade out in large specimens.

GRAHAM'S CRAYFISH SNAKE (*Regina grahami*). Has 158–173 ventrals, 54–67 caudals. Scales in lower row smooth. Belly greenish white to yellow, usually with a narrow median dark line or row of dots. Has a black line where ventral scales meet back scales.

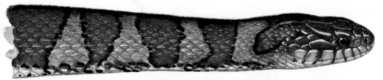

NORTHERN WATER SNAKE
53 in. (135 cm) total

N. f. fasciata

SOUTHERN WATER SNAKE
62½ in. (159 cm) total

N. f. clarki   underside   *N. f. compressicauda*   underside   *N. f. fasciata*

N. f. compressicauda

HARTER'S WATER SNAKE
35½ in. (902 mm) total

underside

KIRTLAND'S SNAKE
24½ in. (622 mm) total

underside

GRAHAM'S CRAYFISH SNAKE
47 in. (121 cm) total

underside

157

QUEEN SNAKE (*Regina septemvittata*). Has 140–153 ventrals, 68–85 caudals. Scales in lower rows are keeled. Belly has 4 rows of dark spots.

GLOSSY CRAYFISH SNAKE (*Regina rigida*). Has 124–142 ventrals, 50–71 caudals. Scales in lower rows are smooth, sometimes with light line. Belly is light, with 2 rows of dark spots joined anteriorly. Three subspecies: (1) *R. r. rigida*—underside of head and sides of throat with dim dark lines (unique); (2) *R. r. deltae*—1 preocular on at least 1 side; (3) *R. r. sinicola*—55 or more caudals in females, 63 or more in males.

STRIPED SWAMP SNAKES—genus *Liodytes*

STRIPED SWAMP SNAKE (*Liodytes alleni;* classified in *Regina* by some). Has 19 rows of scales, smooth except near anus and on tail; 110–133 ventrals; 49–64 caudals; anal divided; 1 internasal, contacting nasals; 1 preocular; 3–4 postoculars; 8 upper labials; 1 anterior temporal.

BLACK SWAMP SNAKES—genus *Seminatrix*

BLACK SWAMP SNAKE (*Seminatrix pygaea*). Has 17 rows of scales, smooth except on tail; anal divided; 112–134 ventrals; 35–56 caudals; single preocular; 2 postoculars; 1 anterior temporal; 8 upper labials. Three subspecies: (1) *S. p. pygaea*—118–124 ventrals; (2) *S. p. cyclas*—112–117 ventrals; (3) *S. p. paludis*—126–134 ventrals.

BROWN SNAKES—genus *Storeria*

Both species in U.S. have keeled scales, divided anal, no loreal, 2 postoculars, 1 anterior temporal. The Brown Snake has 17 scale rows (except 1 subspecies), 7 upper labials, 1 preocular, no light spots, and no collar. The Redbelly has 15 scale rows, 6 upper labials, 2 preoculars, and a pair of light spots or a collar (except 1 subspecies).

BROWN SNAKE (*Storeria dekayi*). Has 112–143 ventrals, 36–63 caudals. Young have light neck ring. Five subspecies: (1) *S. d. dekayi*—paramedian dark spots, small dark blotch on sides of neck, vertical streak near corner of mouth; (2) *S. d. limnetes*—like (1) but longitudinal streak through temporal, no marks on upper labials; (3) *S. d. texana*—like (1) but large dark blotch on sides of neck, no mark on temporal; (4) *S. d. wrightorum*—like (1) but dark spots connected crosswise; (5) *S. d. victa*—15 rows of scales (17 in others).

158

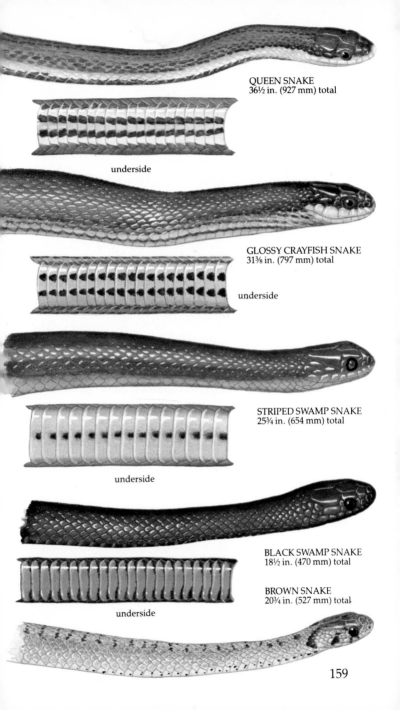

QUEEN SNAKE
36½ in. (927 mm) total

underside

GLOSSY CRAYFISH SNAKE
31⅜ in. (797 mm) total

underside

STRIPED SWAMP SNAKE
25¾ in. (654 mm) total

underside

BLACK SWAMP SNAKE
18½ in. (470 mm) total

BROWN SNAKE
20¾ in. (527 mm) total

underside

159

REDBELLY SNAKE (*Storeria occipitomaculata*). Has 110–133 ventrals, 35–61 caudals. Usually brown, sometimes gray, rarely black. Belly usually red, sometimes yellow to orange, rarely blue-black. Three U.S. subspecies: (1) *S. o. occipitomaculata*—3 separate, well-defined nape spots, light spot on 5th upper labial, black-bordered below; (2) *S. o. obscursa*—spots fused as collar, light spot not black-bordered; (3) *S. o. pahasapae*—spots reduced or absent, no light spot.

COLUBRINE SNAKES—subfamily Colubrinae

In the U.S., 31 genera and 61 species, in 6 groups. All oviparous.

1. BURROWER GROUP

Mostly small and nocturnal, these snakes have compact bodies, and are flatheaded or coneheaded. There are three subgroups.

Terrestrial Burrower Subgroup: Three genera and three species, characterized by smooth scales and a single anterior temporal.

RINGNECK SNAKE (*Diadophis punctatus*). Has 15–17 rows of scales, 111–239 ventrals, 30–76 caudals, divided anal, 2–3 oculars, and 7–8 upper labials. Usually has a reddish neck ring. Twelve subspecies in U.S.

EASTERN (all with belly color usually confined to ventrals): (1) *D. p. punctatus*—1 black spot on each ventral, forming row; (2) *D. p. edwardsi*—few or no small black dots on belly; (3) *D. p. stictogenys*—2 rows of irregular dots on belly, neck ring narrow, often interrupted; (4) *D. p. acricus*—neck ring vestigial or absent; (5) *D. p. arnyi*—belly spots numerous and irregular, usually 17 anterior scale rows (15 in previous 4 subspecies).

WESTERN (all with belly color onto 1 or more dorsal scale rows): (1) *D. p. regalis*—belly color onto 1st dorsal row of scales, 17 scale rows, belly spots numerous, often lacks neck ring, over 200 ventrals in males, 220 in females; (2) *D. p. vandenburghi*—few spots on belly, 17 scale rows, belly color onto 2nd dorsal scale row; (3) *D. p. modestus*—numerous belly spots, always a neck ring, 17 scale rows, fewer ventrals than *D. p. regalis*; (4) *D. p. occidentalis*—15 scale rows, few spots, belly color onto 2nd scale row, dark flecked; (5) *D. p. pulchellus*—like *D. p. occidentalis* but belly color not dark-flecked; (6) *D. p. amabilis*—numerous belly spots, 15 scale rows, belly color onto 2nd row of scales; (7) *D. p. similis*—like *D. p. amabilis* but belly color on no more than lower two-thirds of lower row of scales.

REDBELLY SNAKE
16 in. (406 mm) total

underside

RINGNECK SNAKE
30 in. (760 mm) total

D. p. edwardsi

D. p. edwardsi   underside

D. p. punctatus   underside

D. p. regalis

D. p. amabilis   underside

161

WORM SNAKE (*Carphophis amoenus*). Has 13 rows of scales at midbody, 112–146 ventrals, 23–28 caudals, divided anal, 5 upper labials, no preocular, 1 postocular. Three subspecies: (1) *C. a. amoenus*—brown above, pink of belly onto 1 or 2 lower rows of scales, internasals and prefrontals separate; (2) *C. a. helenae*—like (1) but internasal fused with prefrontals; (3) *C. a. vermis*—purplish black above, pink of belly onto lower 3 rows of scales, internasals and prefrontals separate.

SHARPTAIL SNAKE (*Contia tenuis*). Has 15 rows of scales at midbody, 147–186 ventrals, 25–57 caudals, divided anal; 7 (occasionally 6) upper labials, 1 preocular, 2 postoculars. Reddish or gray. Tail ends in spine, black bar on each ventral.

Aquatic Burrower Subgroup: One genus and two species comprise this subgroup, characterized by 19 scale rows that are smooth (keeled near anus), shiny, and iridescent. Anal usually divided; 6–8 upper labials, 3rd and 4th touching eye. One anterior temporal but no preocular. Body solid and cylindrical; short tail ends in spine.

RAINBOW SNAKE (*Farancia erytrogramma*). Has 158–182 ventrals, 37–49 caudals, 2 internasals. Two subspecies: (1) *F. e. erytrogramma*—2 or 3 large black spots per ventral scale; (2) *F. e. seminola*—belly and lower row of scales black.

MUD SNAKE (*Farancia abacura*). Has 168–208 ventrals, 31–55 caudals, 1 internasal. Two subspecies: (1) *F. a. abacura*—53 or more red bars, pointed at upper ends; (2) *F. a. reinwardti*—52 or fewer, rounded or flat at ends and not extending as far dorsally.

Prowling Burrower Subgroup: There are two genera in this subgroup. Longnose snakes are intermediate between burrower and constrictor groups. Most caudals are undivided (unique in colubrids); the snout is pointed, with the rostral narrow, high, and somewhat protuberant. Scales, in 23–25 rows, are smooth; 1 preocular, 2 postoculars, 1 loreal, usually 8 upper labials, 2 anterior temporals, 181–218 ventrals, 41–61 caudals, and anal undivided. Leafnose snakes are intermediate between burrower and toad-eater groups. Rostral is enlarged, protuberant, triangular, and patchlike, separating the internasals. Both smooth and keeled scales are in 19 rows, with 7–9 preoculars, suboculars, and postoculars; 2–4 loreals, usually 6 upper labials; 2–4 anterior temporals; anal undivided; and a pupil that is vertically oval, not slitlike.

WORM SNAKE
14¾ in. (375 mm) total

C. a. amoenus

C. a. vermis

SHARPTAIL SNAKE
19 in. (483 mm) total

underside

tail spine

RAINBOW SNAKE
66 in. (168 cm) total

underside

MUD SNAKE
81 in. (206 cm) total

underside

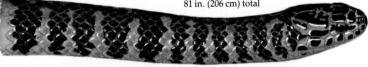

underside

163

LONGNOSE SNAKE (*Rhinocheilus lecontei*). Two subspecies in U.S.: (1) *R. l. lecontei*—snout blunt (no upward tilt), rostral slightly or not raised above adjacent scales, 2 color morphs, 1 banded and the other not; (2) *R. l. tessellatus*—snout sharp (distinct upward tilt), rostral distinctly raised.

SPOTTED LEAFNOSE SNAKE (*Phyllorhynchus decurtatus*). Dorsal blotches 17 or more. Two subspecies in U.S.: (1) *P. d. nubilis*—42–60 middorsal blotches, 157–162 ventrals and 30–33 caudals in males, 171–176 and 20–24 in females; (2) *P. d. perkinsi*—24–48 blotches, 168–182 ventrals and 32–41 caudals in males, 181–196 and 24–34 in females.

SADDLED LEAFNOSE SNAKE (*Phyllorhynchus browni*). Dorsal blotches 16 or fewer. Two subspecies in U.S.: (1) *P. b. browni*—blotches wider than interspaces, 154–170 ventrals and 19–35 caudals in males, 168–178 and 18–24 in females; (2) *P. b. lucidus*—interspaces as wide as or wider than blotches, 166–187 ventrals and 25–40 caudals in males, 176–184 and 29–33 in females.

2. TOAD-EATER GROUP

Hognose snakes (*Heterodon*) are the only U.S. representatives of this group, which is widely scattered over the world. They are characterized by one strongly enlarged tooth, not grooved or hollow, at the rear of each upper jaw. The upper jaws are extensively movable, much as in vipers. Specialized as toad-eaters but sometimes accept other food. Large rear teeth puncture the lungs and deflate toads to swallowable size. These snakes counteract toads' powerful skin poisons by secretion of large amounts of hormones from excessively enlarged adrenal glands. Distinctive behavior when first encountered consists of hissing, lunging with an open mouth, flattening the body, and ultimately playing possum by turning onto the back and becoming limp and seemingly lifeless. They are nocturnal or crepuscular, but have round pupils. They have 23–27 rows of keeled scales; a divided anal; 7–8 upper labials; 8–12 preoculars, suboculars, and postoculars; one or more small azygous (unpaired) scales between the internasals and prefrontals; a rostral that is much enlarged, keeled, rounded, and scooplike; and 2–4 anterior temporals. They prefer sandy soil.

There are 3 species in the U.S. Eastern's upper rostral surface is straight as seen from side; it has 1 azygous scale; underside of tail is lighter than belly. Western and Southern have upper rostral surface turned up, several azygous scales, underside of tail marked like belly. Southern has 25–27 rows of scales; belly is clouded but light. Western has 23 rows of scales and black-checkered belly.

LONGNOSE SNAKE
41 in. (104 cm) total

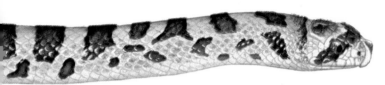

P. d. perkinsi

SPOTTED LEAFNOSE SNAKE
20 in. (508 mm) total

SADDLED LEAFNOSE SNAKE
20 in. (508 mm) total

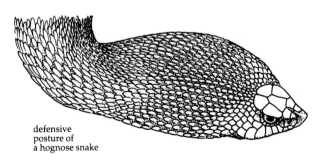

defensive
posture of
a hognose snake

EASTERN HOGNOSE SNAKE (*Heterodon platyrhinos*). Has 25–27 rows of scales, 114–150 ventrals, 37–60 caudals, usually 20–30 large median dark blotches.

SOUTHERN HOGNOSE SNAKE (*Heterodon simus*). Has 115–150 ventrals, 32–55 caudals. There are about 35 dorsal blotches on body, 1 row of same size on each side.

WESTERN HOGNOSE SNAKE (*Heterodon nasicus*). Has 125–152 ventrals, 27–47 caudals, 2 distinct rows of dark spots on each side of median row. Three subspecies: (1) *H. n. nasicus*—9 or more azygous scales, 40–55 blotches at midbody in females, 35 or more in males; (2) *H. n. gloydi*—like (1) but 37 blotches in females, 24-32 in males; (3) *H. n. kennerlyi*—like (2) but 2–6 azygous scales.

3. SIDE-GROOVED, REAR-FANGED GROUP

This is a group of six genera of small, nocturnal snakes with smooth scales in 13-17 rows at midbody, divided caudals and anal, 1 anterior temporal, round pupils, and a head modified for burrowing. Most distinctive (seen only in dead specimens) is a shallow groove on outer side of several rear teeth in upper jaw, sometimes enlarged and/or separated by a short space from the unmodified anterior teeth. Saliva possibly toxic but believed (lacking recorded bites) innocuous to people. Hemipenes (male copulatory organs) are present in females of the Western Hooknose Snake and in several species of this peculiar group not occurring in the U.S. Another anomaly—several species of *Tantilla* lack a left oviduct.

GROUND SNAKES—genus *Sonora*

This is the least specialized genus of the group. The head is only slightly flattened and head scales are normal, with one preocular, two postoculars, and seven upper labials. There are 13–15 rows of scales at the midbody, 126–185 ventrals, and 31–59 divided caudals. The base (anterior edge) of each scale is black.

GROUND SNAKE (*Sonora episcopa*). Has 134–162 ventrals, 31–52 caudals. Pale yellow-brown to orange-red above, black-barred (10–25 bars that are shorter than interspaces), black-collared, or scales streaked. Head may be all black, or with large black spot, or light. Tail is sometimes weakly barred. This species has now been synonymized with *Sonora semiannulata* (Western Ground Snake); see pp. 168–169.

dark color phase

EASTERN HOGNOSE SNAKE
45½ in. (116 cm) total

light color phase

SOUTHERN HOGNOSE SNAKE
24 in. (610 mm) total

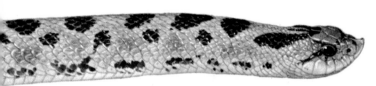

H. n. kennerlyi

WESTERN HOGNOSE SNAKE
35¼ in. (895 mm) total

167

WESTERN GROUND SNAKE (*Sonora semiannulata*). Usually has 15 rows of scales at midbody, 13 or 14 rows at rear; 149–185 ventrals, 40–60 caudals. Solid yellow-brown, or with broad vertebral reddish stripe that is combined with or replaced by 19–40 black bars, encircling body or limited to back. Great variation occurs in color and pattern within and between populations. Some are uniformly yellow-brown or orange. Some have a single, diffuse vertebral stripe that can be reddish orange or tan. Others have 19–40 black bars or "saddles" that encircle the body or are limited to the back. These banded individuals often have orange or red interspaces. Some have a single black neck band.

SHOVELNOSE SNAKES—genus *Chionactis*

Flatheaded, shovelnosed derivatives of *Sonora*–like ancestors, these have 15 rows of smooth scales throughout, with 141–181 ventrals and 34–64 caudals; a snout projecting beyond the lower jaw; a nasal valve; normal head scales; one preocular; two postoculars; seven upper labials. The belly is angular. The Sonoran Shovelnose has a snout that is convex above, usually fewer than 21 bars, and a black crescent that is straight in section between eyes. The Western has a flat snout, 21 or more dark bars, and a black crescent that is concave in section between eyes.

WESTERN SHOVELNOSE SNAKE (*Chionactis occipitalis*). Has 16–40 dark rings or crossbars. Five subspecies: (1) *C. o. occipitalis*—no black-brown secondary bars in interspaces, primary bars on body plus anterior bars not reaching ventrals usually 45 bars or more; (2) *C. o. annulata*—like (1) but usually 52 or fewer caudals in males, 47 or fewer in females; (3) *C. o. saxatilis*—like (2) but usually 53 or more caudals in males, 48 or more in females; (4) *C. o. klauberi*—distinct black-brown secondary bars in interspaces, 42 or fewer caudals, usually 148 or fewer ventrals in males, 158 or fewer in females; (5) *C. o. talpina*—like (4) but 43 or more caudals, 149 or more ventrals in males, 159 or more in females.

SONORAN SHOVELNOSE SNAKE (*Chionactis palarostris*). Has 10–20 dark rings or crossbars on body, most completely across belly, bordered on each side by yellow ring half as long as black rings; triads of yellow-black-yellow separated by red bars. One subspecies in U.S.: *C. p. organica*.

168

WESTERN GROUND SNAKE
19 in. (483 mm) total

color variations

WESTERN SHOVELNOSE SNAKE
17¾ in. (451 mm) total

SONORAN SHOVELNOSE SNAKE
15⅜ in. (391 mm) total

169

BANDED SAND SNAKES—genus *Chilomeniscus*

BANDED SAND SNAKE (*Chilomeniscus cinctus*). Has 13 rows of scales, 108–125 ventrals, 21–29 caudals, angular belly, 18–32 black crossbars, 7 upper labials, usually no loreal. Flattened snout extends far beyond countersunk lower jaw, rounded in dorsal view, pointed in lateral profile. Internasals fused with anterior part of nasal scale. Has nasal valve (greatly enlarged), rostral often reaching prefrontals, 1 preocular, 2 postoculars.

FLATHEAD SNAKES—genus *Tantilla*

About 40 species, extending from U.S. (10 species) to Argentina. All have 15 scale rows, no loreal, 1 preocular, a flat-topped head, and 1 anterior temporal. Most have 7 upper labials, and 2 postoculars.

KEY TO FLATHEAD SNAKES

1. Head scarcely if any darker than body . . . . **Flathead,** p. 174
   Head black above . . . . *see* 2
2. E of Mississippi River . . . . . *see* 3
   W of Mississippi River . . . . *see* 5
3. Collar with front border at rear level of 7th labial, postocular labials mostly light . . . .
   .**Southeastern Crowned,** below
   Not as above . . . . *see* 4
4. No complete collar, 2 notches at rear sides of cap . . . .
   . . . **Rim Rock Crowned,** below
   If no collar, 1 notch; if collared, front border at front level of 7th labial, mostly dark postocular labials . . . . **Florida,** p. 172

5. Head all black, or collar bordered behind by black bar 3 scales long or longer . . . **Big Bend,** below
   Not as above . . . . *see* 6
6. Nuchal collar, rear broadly black-edged . . . **Chihuahuan,** p. 172
   Not as above . . . . *see* 7
7. Rear edge of head cap V-shaped .
   . . . . .**Plains Blackhead,** p. 174
   Not as above . . . . *see* 8
8. Temporal-labial white spot . . . . .
   . . . . . **Yaqui Blackhead,** p. 172
   None . . . . *see* 9
9. Ventrals 135–151 in males, 145–160 in females . . . . **Mexican,** p. 172
   More . . . . **Western,** p. 172

BIG BEND BLACKHEAD SNAKE (*Tantilla rubra*). Has 170–181 ventrals, 63–73 caudals. Light brown above, belly whitish. Two subspecies in U.S.: (1) *T. r. diabola*—black head cap, white collar followed by broad (3.5–4.5 scales) black bar, light spot on upper lip behind eye; (2) *T. r. cucullata*—head all black.

SOUTHEASTERN CROWNED SNAKE (*Tantilla coronata*). Has 123–147 ventrals, 34–45 caudals, 2 basal hooks on hemipenis. Tan to dark brown, with white belly.

RIM ROCK CROWNED SNAKE (*Tantilla oolitica*). Has 135–146 ventrals, 45–63 caudals, 2 basal hooks on hemipenis. Tan, with white belly.

BANDED SAND SNAKE
10 in. (254 mm) total

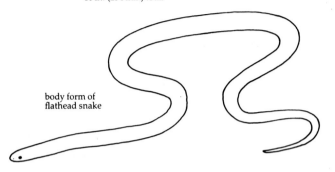

body form of
flathead snake

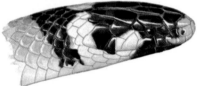

T. r. diabola

BIG BEND BLACKHEAD SNAKE
21¾ in. (552 mm) total

SOUTHEASTERN CROWNED SNAKE
13 in. (330 mm) total

RIM ROCK CROWNED SNAKE
11½ in. (292 mm) total

FLORIDA CROWNED SNAKE (*Tantilla relicta*). Has 1 basal hook on hemipenis. Belly is pinkish gray. Three subspecies: (1) *T. r. relicta*—light tan, light collar, 117–131 ventrals and 44–59 caudals in males, 120–134 ventrals and 40–55 caudals in females; (2) *T. r. pamlica*—reddish brown above, light collar, 45–51 caudals, 115–118 ventrals in males, 119–121 ventrals in females, tendency for loss of pigment on parietals, temporals, and/or snout (not so in 1); (3) *T. r. neilli*—dark tan, no definite light collar, 123–135 ventrals and 51–67 caudals in males, 129–142 ventrals and 46–60 caudals in females.

CHIHUAHUAN BLACKHEAD SNAKE (*Tantilla wilcoxi*). Has 149–164 ventrals, 62–69 caudals. Light collar covers 2 scale lengths and tips of parietals. Belly red in middle and rear, unmarked with black. One subspecies in U.S.: *T. w. wilcoxi*.

WESTERN BLACKHEAD SNAKE (*Tantilla planiceps*). Has 47–73 caudals. Yellowish brown to dark olive-gray above, belly white with reddish streak. Three subspecies in U.S.: (1) *T. p. eiseni*—head cap 1.5–2.5 scale lengths onto nape and to or below angle of mouth, collar distinct, 163–175 ventrals in males, 167–185 in females; (2) *T. p. transmontana*—like (1) but ventrals 175–185 in males, 187–198 in females; (3) *T. p. utahensis*—head cap less than 1.5 scale lengths onto nape, not to angle of mouth, collar indistinct or absent, ventrals 153–165 in males, 162–174 in females.

MEXICAN BLACKHEAD SNAKE (*Tantilla atriceps*). Has 135–151 ventrals in males, 145–160 in females. Head cap less than 1.5 scale lengths onto nape, not to angle of mouth, collar faint or absent. Dark pigment on upper edge of upper labials.

YAQUI BLACKHEAD SNAKE (*Tantilla yaquia*). Has 46–75 caudals; 134–157 ventrals in males, 145–165 in females. White labial-temporal patch (no dark pigment on upper labials 5 and 6). Head cap 2.5–4 scale lengths onto nape, to or below angle of mouth, collar distinct. Belly pinkish orange, brighter posteriorly.

FLORIDA CROWNED SNAKE
9½ in. (241 mm) total

CHIHUAHUAN BLACKHEAD SNAKE
13¾ in. (349 mm) total

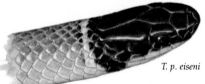

T. p. eiseni

WESTERN BLACKHEAD SNAKE
15 in. (381 mm) total

MEXICAN BLACKHEAD SNAKE
9⅝ in. (244 mm) total

YAQUI BLACKHEAD SNAKE
12¾ in. (324 mm) total

173

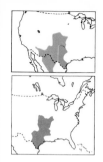

PLAINS BLACKHEAD SNAKE (*Tantilla nigriceps*). Has 35–62 caudals. Dark head cap, usually pointed, 2–5 scale lengths behind parietals. No light collar, belly whitish with reddish streak. Two subspecies: (1) *T. n. nigriceps*—146–161 ventrals; (2) *T. n. fumiceps*—130–145 ventrals.

FLATHEAD SNAKE (*Tantilla gracilis*). Has 106–138 ventrals, 33–56 caudals, 1 postocular, 6 upper labials. Yellowish to reddish brown above, head darker but rarely distinctly black capped. Belly white with reddish streak.

HOOKNOSE SNAKES—genera *Gyalopion* and *Ficimia*

These are small burrowers with smooth scales in 17 rows. The rostral is enlarged, with a turned-up tip. Also present are one preocular, two postoculars, and one anterior temporal. The two species of *Gyalopion* are both in the United States; of the six of *Ficimia*, one is in the United States. In the Mexican Hooknose, the rostral contacts the frontal; in others, they are separated by prefrontals. The anal is divided in the Western and Mexican, entire in the Desert. Snout is flattened and concave behind hook; hognose snakes have a middorsal keel on the snout.

WESTERN HOOKNOSE SNAKE (*Gyalopion canum*). Has 122–140 ventrals, 26–37 caudals. Internasals and loreal are present, and 7 upper labials. Pale brown, with 25–48 dark brown blotches longer than interspaces; belly white or cream. Often everts (turns inside out) cloacal lining with a bubbling or popping sound when captured.

DESERT HOOKNOSE SNAKE (*Gyalopion quadrangularis*). Has 116–140 ventrals, 20–32 caudals. Internasals are present; loreal absent in northern areas, present in southern. Has 5–7 (usually 6) upper labials. Black blotches (16–41 on body) are shorter than interspaces, red streaks on sides are interrupted by blotches. Ground color is pale gray, belly is white.

MEXICAN HOOKNOSE SNAKE (*Ficimia streckeri*). Has 126–155 ventrals, 28–41 caudals. No internasal or loreals, but 7 upper labials. Brown-gray, with 36–60 irregular, narrow darker brown or olive crossbars on body, which may be broken into spots. Belly is pale yellowish.

PLAINS BLACKHEAD SNAKE
14¾ in. (375 mm) total

FLATHEAD SNAKE
9⅝ in. (244 mm) total

WESTERN HOOKNOSE SNAKE
14⅜ in. (365 mm) total

DESERT HOOKNOSE SNAKE
14 in. (356 mm) total

MEXICAN HOOKNOSE SNAKE
18⅞ in. (479 mm) total

4. FRONT-GROOVED, REAR-FANGED GROUP

This group includes five genera, each with one species in North America. Lyre, Black-striped, and Cat-eyed snakes have a deep groove down front margin of the two rear teeth in upper jaw, enlarged and usually preceded by a space. All have weak venom that has little effect on humans. Lyre, Cat-eyed, and Night snakes have vertical pupils; pupils are round in others. Scales are smooth, in 17–27 rows; anal is divided (1 exception). There are paired caudals, 2 or more preoculars (2 exceptions) and postoculars, and 7–9 upper labials.

PINE WOODS SNAKE (*Rhadinaea flavilata*). Has 17 rows of scales, 112–139 ventrals, 59–83 caudals, normal head scales, 1 preocular, usually 7 (occasionally 8) upper labials, 1 anterior temporal.

BLACK-STRIPED SNAKE (*Coniophanes imperialis*). Has 19 rows of scales, 127–140 ventrals, 68–78 caudals, normal head scales, usually 1 (occasionally 2) preoculars, 8 upper labials, 1 anterior temporal. One subspecies: *C. i. imperialis*.

NIGHT SNAKE (*Hypsiglena torquata*). Has 162–204 ventrals, 38–66 caudals, normal head scales, 1 anterior temporal. Six subspecies: (1) *H. t. nuchalata*—19 rows of scales (21 in others); (2) *H. t. lorealis*—2 loreals (usually 1 in others); (3) *H. t. klauberi*—median nape spot usually extending to 3 scales behind parietal (farther in others); (4) *H. t. deserticola*—spot at least 3 times as wide at rear as at front (less in others); (5) *H. t. texana*—dorsal spots large, involving 22 or more scales; (6) *H. t. ochrorhyncha*—spots smaller.

CAT-EYED SNAKE (*Leptodeira septentrionalis*). Has 23 (occasionally 21) rows of scales at midbody, 181–208 ventrals, 60–94 caudals, 3 preoculars, 8 upper labials, 1 anterior temporal. Head is broad, flat, wider than neck.

LYRE SNAKE (*Trimorphodon biscutatus*). Has 21–24 rows of scales, 220–244 ventrals, 58–86 caudals. Head is wider than neck, and eyes are protuberant. Usually has 3 preoculars and 3 postoculars; 2–5 loreals, 8–10 upper labials, 2 anterior temporals. Three subspecies: (1) *T. b. vilkinsoni*—no **V** mark on top of head, 17–24 blotches, anal divided; (2) *T. b. vandenburghi*—distinct **V**, 28–43 long blotches, each split transversely, anal usually entire; (3) *T. b. lambda*—like (2) except 21–34 blotches, anal divided.

PINE WOODS SNAKE
15⅞ in. (403 mm) total

BLACK-STRIPED SNAKE
20 in. (508 mm) total

C. i. imperialis

NIGHT SNAKE
26 in. (660 mm) total

H. t. deserticola

CAT-EYED SNAKE
38⅞ in. (985 mm) total

T. b. lambda

LYRE SNAKE
47¾ in. (121 cm) total

T. b. vilkinsoni

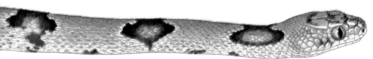

5. CONSTRICTOR GROUP

Six genera in the U.S. They constrict their prey, feeding mainly on birds and mammals; a few feed on snakes and lizards. All appear to have evolved from the burrower group. Though basically nocturnal, there is a trend toward diurnal activity in advanced members. Primitive members have smooth scales, the more advanced (*Elaphe, Pituophis*) have keeled scales. All except *Elaphe* have an entire anal. Except for the most primitive genera (*Stilosoma* and *Cemophora*), all are relatively large.

SHORT-TAILED SNAKE (*Stilosoma extenuatum*). Has 19 rows of scales, 223–260 ventrals, 33–46 caudals. Body is slender; internasal is often fused with prefrontals. There is no loreal or anterior temporal; prefrontals and parietals contact; there are 6 upper labials.

SCARLET SNAKE (*Cemophora coccinea*). Has 19 rows of scales, 149–195 ventrals, 31–50 caudals. Rostral is enlarged, wider than high, extending over half the distance to frontal. Head scales are normal; has 1 loreal, 1 preocular, 1 postocular, 1 anterior temporal. Three subspecies: (1) *C. c. coccinea*—black borders enclosing red blotches laterally, 7 supralabials, 1st black border separated from parietals by 2 or more scale lengths; (2) *C. c. copei*—black borders as in (1), 6 supralabials, black on neck closer to parietals; (3) *C. c. lineri*—like (1) but black borders not enclosing.

KINGSNAKES—genus *Lampropeltis*

Six species occur in the U.S., all with smooth scales, an entire anal, divided caudals, 17–27 rows of scales at midbody, 7 upper labials, 1 preocular, 2–3 postoculars, 1 loreal, usually 2 anterior temporals, and a cylindrical head. Kingsnakes are docile after capture.

KEY TO KINGSNAKES

1. Distinct red or orange . . . . . *see* 2
   Without red or orange . . . . *see* 5
2. Red blotches narrower than interspaces, or fewer than 15 . . . . . .
   . . . .**Gray-banded,** p. 182
   Blotches wider, 17 or more . *see* 3
3. W Coast states . . . .
   . . **California Mountain,** p. 182
   Not as above . . . . *see* 4
4. 37–61 yellow or white rings. . . . .
   . . . **Sonoran Mountain,** p. 182
   Fewer rings . .**Milk Snake,** p. 180

5. 3-toned above (black, brownish, whitish) and dorsal blotches reach at least 5th row of scales . . . . .**Milk Snake,** p. 180
   Not as above . . . . *see* 6
6. Unicolor, dull-striped or dorsal blotches not reaching 5th scale rows; scales not light-centered . . . . **Prairie,** p. 180
   Brightly striped or scales light-centered or dorsal blotches extending below 5th scale row . . .
   . . **Common Kingsnake,** p. 180

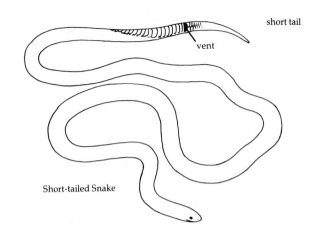

short tail

vent

Short-tailed Snake

SHORT-TAILED SNAKE
25¾ in. (654 mm) total

SCARLET SNAKE
32½ in. (826 mm) total

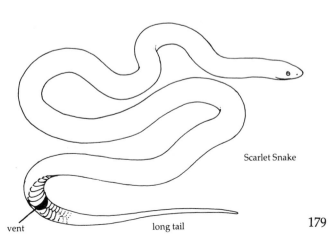

Scarlet Snake

vent

long tail

179

COMMON KINGSNAKE (*Lampropeltis getulus*). Has 199–254 ventrals, 38–62 caudals, usually 21–23 scale rows at midbody. Seven subspecies in U.S.: (1) *L. g. getulus*—black with 23–52 narrow white or yellow crossbands, belly checkered; (2) *L. g. floridana*—brownish, scales yellowish at base, 46–85 light crossbands; (3) *L. g. niger*—black with chainlike pattern; (4) *L. g. holbrooki*—adults with each scale center light, young crossbanded; (5) *L. g. splendida*—lateral scales light-centered, belly black, dark dorsal blotches; (6) *L. g. californiae*—broad light bands, sometimes dark blotches fused as stripes, light vertebral line; (7) *L. g. nigrita*—all black, little or no light markings.

PRAIRIE KINGSNAKE (*Lampropeltis calligaster*). Has 191–215 ventrals; 31–57 caudals; 46–78 blotches, sometimes dim or with dull stripes. Two subspecies: (1) *L. c. calligaster*—25–27 rows of scales, median spots with concave margins, upper lateral spots round; (2) *L. c. rhombomaculata*—21–23 rows of scales, median spots with convex margins, upper lateral spots elongate vertically.

MILK SNAKE (*Lampropeltis triangulum*). Has 19–23 (usually 21) scale rows, 152–215 ventrals, 29–57 caudals. Nine subspecies in U.S.

EASTERN: (1) *L. t. triangulum*—marks on nape longitudinal (transverse in all others), 27–51 blotches, brown in adult (red in young and in all other subspecies) usually to 5th (sometimes 2nd) scale row, 2 lateral rows of small spots (1 or none in all others); (2) *L. t. syspila*—16–31 blotches to 1st and 2nd scale rows, alternate with small lateral blotches; (3) *L. t. amaura*—13–25 red blotches or rings, no lateral blotches; (4) *L. t. elapsoides*—13–19 red rings separated by black-yellow-black triads, head red and unspotted (spotted or black in all others), 19 scale rows, 152–193 ventrals (usually more than 187 in all others).

WESTERN: (1) *L. t. gentilis*—20–32 reddish rings, black expanded across red middorsally and midventrally; (2) *L. t. multistrata*—22–32 orange rings, belly mostly white, few dark marks; (3) *L. t. celaenops*—14–26 red rings complete ventrally, white rings expanded middorsally; (4) *L. t. taylori*—24–40 red rings complete, black rings incomplete ventrally; (5) *L. t. annulata*—11–20 red rings, snout and belly mostly black.

180

L. g. getulus

COMMON KINGSNAKE
82 in. (208 cm) total

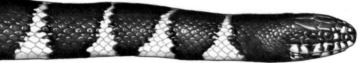

L. g. californiae

PRAIRIE KINGSNAKE
53⅜ in. (136 cm) total

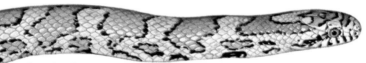

L. t. triangulum

MILK SNAKE
52 in. (132 cm) total

L. t. syspila

L. t. elapsoides

181

GRAY-BANDED KINGSNAKE (*Lampropeltis mexicana*). Has 23–27 rows of scales at midbody, 19 or more at rear (17 in most other kingsnakes); 208–229 ventrals (in U.S.); 55–67 caudals. Markings vary from narrow black rings with no red to broad blotches, more red than black. Belly is mostly white, black-blotched, or mostly black. One subspecies in U.S.: *L. m. alterna*.

SONORAN MOUNTAIN KINGSNAKE (*Lampropeltis pyromelana*). Has 23–25 rows of scales, 216–235 ventrals, 61–79 caudals. Three subspecies in N.A.: (1) *L. p. pyromelana*—10 lower labials, usually 43–61 white rings, fewer than half complete across belly; (2) *L. p. woodini*—like (1) but usually 37–42 rings; (3) *L. p. infralabialis*—9 lower labials, 42–57 white belly rings, half or more complete.

CALIFORNIA MOUNTAIN KINGSNAKE (*Lampropeltis zonata*). Has 21–25 (usually 23) rows of scales at midbody, 194–227 ventrals, 46–62 caudals. Five U.S. subspecies: (1) *L. z. zonata*—1st white ring not on rear upper labial, 40% or fewer triads united at midline; (2) *L. z. multicincta*—like (1) but more than 40% of triads united; (3) *L. z. parvirubra*—1st white ring on rear upper labial, 37 or more triads; (4) *L. z. pulchra*—like (3) but 36 or fewer triads; (5) *L. z. multifasciata*—snout red-flecked (black in others).

GLOSSY SNAKES—genus *Arizona*

GLOSSY SNAKE (*Arizona elegans*). Has 25–31 rows of scales at midbody, 185–241 ventrals, 39–63 caudals. Rostral is somewhat enlarged; head scales are normal. Usually has 1 preocular; has 2 postoculars, 8 upper labials, 2 anterior temporals. Ten subspecies, 7 in U.S.: (1) *A. e. elegans*—usually 29–31 rows of scales, 51 or more blotches, 211 or fewer ventrals in males, 220 or fewer in females; (2) *A. e. arenicola*—29–31 rows of scales, 50 or fewer blotches, 212 or more ventrals in males, 221 or more in females; (3) *A. e. philipi*—usually 27 rows of scales (also in following 4 subspecies), tail more than 14.5% of total length in males, 13.5% in females (also in preceding 2); (4) *A. e. occidentalis*—tail shorter (also in following 3 subspecies), dark spots on lower lips and edges of ventrals (not so in next 3); (5) *A. e. noctivaga*—length of blotches at midline equals or exceeds length of interspaces (not in next 2); (6) *A. e. eburnata*—blotches 7 scales wide, 1 preocular; (7) *A. e. candida*—blotches 9 scales wide, usually 2 preoculars.

GRAY-BANDED KINGSNAKE
47½ in. (121 cm) total

color variations

SONORAN MOUNTAIN KINGSNAKE
42 in. (107 cm) total

CALIFORNIA MOUNTAIN KINGSNAKE
40 in. (102 cm) total

GLOSSY SNAKE
56 in. (142 cm) total

RAT SNAKES—genus *Elaphe*

The five species in the United States all have 25–39 rows of scales at midbody, at least some weakly keeled; 190–282 ventrals; 45–126 caudals; normal head scales; 1 loreal; 1 preocular; 2 postoculars; 8–12 upper labials; 2 or more anterior temporals; and a broad, flat head.

KEY TO RAT SNAKES

1. Suboculars present; eye usually not contacting labials . . . .
. . . . **Trans-Pecos,** below
Not as above . . . . *see* 2
2. 31 or more rows of scales . . . .
. . . . **Green,** below
Fewer rows . . . . *see* 3

3. Black-edged dark marks on head, inverted **V** with apex between eyes . . . . . . **Corn Snake,** below
Not as above . . . . *see* 4
4. 220 or more ventrals . . . .
. . . . **Rat Snake,** below
216 or fewer . . . **Fox Snake,** p. 186

TRANS-PECOS RAT SNAKE (*Elaphe subocularis*). Has 31–35 rows of scales, 260–277 ventrals, 69–79 caudals, 3–6 suboculars, usually 10–11 upper labials, enlarged eye, and pink tongue. Tail usually lost to tick parasites.

GREEN RAT SNAKE (*Elaphe triaspis*). Has 30–39 rows of scales, 243–282 ventrals, 89–126 caudals, usually 8 upper labials. One subspecies in U.S.: *E. t. intermedia.*

CORN SNAKE (*Elaphe guttata*). Has 27 (occasionally 29) rows of scales, 207–245 ventrals, 60–97 caudals, and 8 upper labials. Two subspecies: (1) *E. g. guttata*—red blotches, black-bordered; (2) *E. g. emoryi*—blotches gray-brown on gray ground.

RAT SNAKE (*Elaphe obsoleta*). Has 21–29 (usually 25 or 27) rows of scales, 221–258 ventrals, 67–103 caudals, 8 upper labials. Young always have blotches, 25–40 in median row, lateral row of smaller spots; no stripe through eye, or head all dark. Six subspecies: (1) *E. o. obsoleta*—adults uniformly black above, or spots dimly evident; (2) *E. o. spiloides*—blotches throughout life, contrasty, 231–258 ventrals; (3) *E. o. lindheimeri*—blotches throughout life, less contrasty, 218–237 ventrals; (4) *E. o. quadrivittata*—adults dark-striped, with yellow ground above, belly whitish, gray-clouded to rear; (5) *E. o. rossalleni*—narrow dark stripes, red-orange ground above and below, tongue red; (6) *E. o. bairdi*—adults dark-striped on gray ground, young with 46–53 blotches, often 9 upper labials, posterior chin shields often split transversely, ventrals and caudals much higher than in adjacent subspecies (may be a distinct species).

TRANS-PECOS RAT SNAKE
66 in. (168 cm) total

GREEN RAT SNAKE
57 in. (145 cm) total

E. g. guttata

CORN SNAKE
72 in. (183 cm) total

E. g. emoryi

E. o. obsoleta

RAT SNAKE
101 in. (257 cm) total

E. o. quadrivittata

FOX SNAKE (*Elaphe vulpina*). Has 23–27 rows of scales, 192–216 ventrals, 53–69 caudals, usually 8 upper labials. Two subspecies: (1) *E. v. vulpina*—32–42 blotches, head yellowish; (2) *E. v. gloydi*—28–39 blotches, head reddish.

BULLSNAKES—genus *Pituophis*

BULLSNAKE (*Pituophis melanoleucus*). Has 27–37 (usually 28–33) rows of strongly keeled scales at midbody; 205–262 ventrals; 47–84 caudals; undivided anal; small, conical head; 7–10 (usually 8) supralabials, only 1 touching eye; usually 1–2 preoculars, 2–4 postoculars; 4 prefrontals; enlarged rostral. Hisses loudly when alarmed but quickly becomes docile. Ten subspecies in U.S.

EASTERN: (1) *P. m. melanoleucus*—rostral narrow, about twice as long as broad, 29 or fewer blotches distinct throughout life; (2) *P. m. mugitus*—like (1) but blotches indistinct in adults; (3) *P. m. lodingi*—like (1) but black or dark brown above and below; (4) *P. m. ruthveni*—rostral 1.5 times as long as wide, head markings reduced or absent, usually 30–40 blotches.

WESTERN: (1) *P. m. sayi*—rostral like *P. m. ruthveni*, head markings distinct, usually more than 40 blotches; (2) *P. m. affinis*—rostral slightly or not longer than broad (also in following 4 subspecies), anterior blotches not connected, cream or buff; (3) *P. m. catenifer*—like *P. m. affinis* but suffused with gray; (4) *P. m. deserticola*—anterior blotches connected at sides, cream or buff; (5) *P. m. annectens*—like *P. m. deserticola* but suffused with gray; (6) *P. m. pumilis*—27–29 scale rows (more in others).

6. DIURNAL GROUP

Seven genera, with sufficiently good vision to be active mainly in the daytime. All have long tails and are adapted for rapid movement. None are constrictors. That all the genera in this group in the U.S. are more closely related to each other than to some other genera is not clearly established. *Opheodrys* and *Oxybelis* have only 1 anterior temporal (all others have 2). Green snakes have 116–154 ventrals (others have no fewer than 142 and as many as 217), and 7 upper labials (others have 8–10). Mexican Vine Snakes have rear fangs, but grooves are deep and lateral as opposed to anterior or shallow in other rear-fanged groups. All except the Indigo have a divided anal; all have 15–17 rows of scales at midbody. Scales are smooth except in *Oxybelis*, *Drymobius*, and one species of *Opheodrys*. Often 2 or more preoculars, 2–4 postoculars, 1 loreal except in *Oxybelis* (none) and *Salvadora* (1–4), and 7–10 upper labials.

FOX SNAKE
70½ in. (179 cm) total

BULLSNAKE
108 in. (275 cm) total

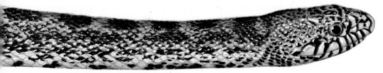

P. m. melanoleucus

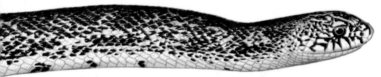

P. m. ruthveni

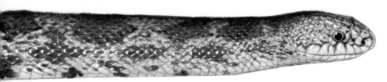

P. m. sayi

P. m. deserticola

GREEN SNAKES—genus *Opheodrys*

SMOOTH GREEN SNAKE (*Opheodrys vernalis*). Has 15 rows of smooth scales; often has 2 preoculars. Two subspecies: (1) *O. v. vernalis*—116–137 (usually fewer than 131) ventrals in males, 121–144 (usually fewer than 139) in females; (2) *O. v. blanchardi*—125–141 (usually more than 131) ventrals in males, 139–154 (usually more than 140) in females.

ROUGH GREEN SNAKE (*Opheodrys aestivus*). Has 17 rows of keeled scales; sometimes has 2 preoculars and 2 anterior temporals; 139–171 ventrals and 114–152 caudals.

MEXICAN VINE SNAKES—genus *Oxybelis*

MEXICAN VINE SNAKE (*Oxybelis aeneus*). Has 17 rows of scales, 173–205 ventrals, no loreal, 1 preocular, 2 postoculars, 1 anterior temporal, 2–4 enlarged scales contacting rear edge of parietals, and 8–10 upper labials. Head and body are extremely elongate.

SPECKLED RACERS—genus *Drymobius*

SPECKLED RACER (*Drymobius margaritiferus*). Has 17 rows of scales, weakly keeled middorsally; 142–168 ventrals; 85–126 caudals; 1 preocular; 2 postoculars; usually 9 (occasionally 8) upper labials. One subspecies in U.S.: *D. m. margaritiferus*.

INDIGO SNAKES—genus *Drymarchon*

INDIGO SNAKE (*Drymarchon corais*). Has 17 rows of scales at midbody; 182–217 ventrals; 55–88 caudals; undivided anal; 1 preocular; 2 postoculars; usually 8 (occasionally 7 and rarely 9) upper labials, 3rd from last distinctively narrowed at top. Subocular labial is black-edged. This is the longest native snake in the U.S. Two subspecies in U.S.: (1) *D. c. couperi*—adults black, labials on either side of 3rd from last contacting above it; (2) *D. c. erebennus*—adults black posteriorly above and below, lighter but irregularly black-flecked and streaked above anteriorly, belly reddish white to midbody, usually 14 scale rows in front of anus, labials on each side of 3rd from last not contacting above.

SMOOTH GREEN SNAKE
26 in. (660 mm) total

ROUGH GREEN SNAKE
45½ in. (116 cm) total

MEXICAN VINE SNAKE
78 in. (198 cm) total

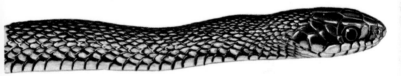

SPECKLED RACER
50 in. (127 cm) total

D. c. couperi

INDIGO SNAKE
116½ in. (295 cm) total

189

RACERS—genus *Coluber*

RACER (*Coluber constrictor*). Usually has 17 rows of scales at midbody; has 158–193 ventrals, 66–119 caudals, 2 preoculars, 2 postoculars, and 2 anterior temporals. Eleven subspecies, distinguishable only as adults.

EASTERN: (1) *C. c. constrictor*—dorsal and most of ventral surfaces black, little white on chin, throat, or lips; (2) *C. c. helvigularis*—like (1) but moderately extensive light areas and light brown on chin, throat, and lips; (3) *C. c. priapus*—like (2) but light areas white; chin, throat, and lips black; (4) *C. c. anthicus*—scattered irregular light patches on ground color of blue-gray-black; (5) *C. c. etheridgei*—like (4) except light tan; (6) *C. c. foxi*—back gray-olive-bluish, belly bluish, iris brownish; (7) *C. c. paludicola*—like (6) except iris reddish-yellowish; (8) *C. c. latrunculus*—slate-gray above, grayish blue below, black postocular stripe (lacking in all others).

WESTERN: (1) *C. c. mormon*—8 upper labials (7 in all others except *C. c. oaxaca*), 9 lower labials, olive above, yellowish below; (2) *C. c. oaxaca*—like *C. c. mormon* except 8 lower labials, greenish above; (3) *C. c. flaviventris*—back gray-olive-bluish, belly yellowish.

WHIPSNAKES—genus *Masticophis*

The four species in the United States are similar to the Racer but have 13 (as opposed to 15) rows of scales at rear of body and average a larger size. Three are striped (the Racer isn't). The Coachwhip has a distinctive color as an adult; its young have narrow crossbars rather than blotches. Head scales are similiar to the Racer's but an elongate bell-shaped frontal is distinctive. They regularly have eight upper labials.

KEY TO WHIPSNAKES

1. 15 rows of scales at midbody . . . .
. . . . **Striped,** p. 192
17 rows . . . . *see* 2
2. Well-defined lateral light stripes
. . . . *see* 3
Not as above . **Coachwhip,** p. 192

3. Single lateral light stripe continuous onto tail . . . .
. . . . . . **Striped Racer,** below
2–3 light stripes on each side, not reaching tail . . . .
. . . . **Sonoran,** p. 192

STRIPED RACER (*Masticophis lateralis*). Has 183–204 ventrals and 115–137 caudals. Two subspecies: (1) *M. l. lateralis*—stripes yellowish, narrower, on adjacent halves of scale rows 3 and 4, cream below, pink under tail; (2) *M. l. euryxanthus*—stripes and anterior belly orange, stripes wider, on scale row 3 and edges of adjacent rows.

190

C. c. constrictor

RACER
75¼ in. (191 cm) total

Racer
juvenile

C. c. foxi

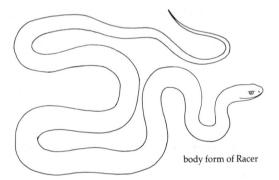

body form of Racer

bell-shaped
frontal of
whipsnake

STRIPED RACER
60 in. (152 cm) total

COACHWHIP (*Masticophis flagellum*). Has 173–212 ventrals, 91–129 caudals. Seven subspecies, best identified by locality, but general characters follow: (1) *M. f. flagellum*—brown to black anteriorly above and below, tannish above and whitish to pale yellowish below posteriorly; (2) *M. f. testaceus*—light brown to pinkish red above, narrow dark crossbands on neck, double row of dark spots on pale yellow belly; (3) *M. f. lineatulus*—light brown-gray above, median dark streak on anterior of each dorsal scale, underside of tail and rear belly salmon-pink; (4) *M. f. piceus*—black above and salmon-pink to red toward rear below, or pink-red all over except for dark crossbands on neck; (5) *M. f. ruddocki*—light yellow to olive-yellow above, no dark crossbands; (6) *M. f. cingulum*—usually wide reddish brown crossbands separated by narrow light interspaces split by narrower dark bar, sometimes uniformly black or reddish brown; (7) *M. f. fuliginosus*—yellow to light gray with dark edges on many scales and black crossbands on neck, yellow below, or dark brown above and lighter to rear with dark brown under head and neck grading to creamy white toward rear.

SONORAN WHIPSNAKE (*Masticophis bilineatus*). Has 187–209 ventrals, 129–149 caudals. Two subspecies: (1) *M. b. bilineatus*—dorsolateral light line wider, more distinct than other lines, on adjacent halves of 3rd and 4th scale rows usually beginning at 4th scale behind rear upper labial; (2) *M. b. lineolatus*—dorsolateral light line narrower, less distinct than others, on mere edges of 3rd and 4th scale rows beginning at 8th scale.

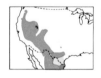

STRIPED WHIPSNAKE (*Masticophis taeniatus*). Has 15 rows of scales at midbody, 188–214 ventrals, 122–160 caudals. Four U.S. subspecies: (1) *M. t. taeniatus*—black to reddish brown above, upper lateral line on 4th and edges of adjacent scale rows, continuous and uniform throughout, reaching anal region; (2) *M. t. girardi*—like (1) but upper lateral light line broadly interrupted at regular intervals and not reaching anal region; (3) *M. t. schotti*—bluish to greenish gray above, sides of neck reddish, upper lateral light line confined to adjacent edges of 3rd and 4th scale rows, lower line on edges of ventrals and 1st scale row; (4) *M. t. ruthveni*—no distinct lines, or if so, they are narrow and confined to neck, belly bright yellow anteriorly.

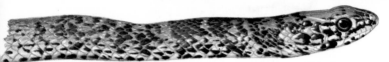

COACHWHIP
102 in. (259 cm) total

M. f. piceus

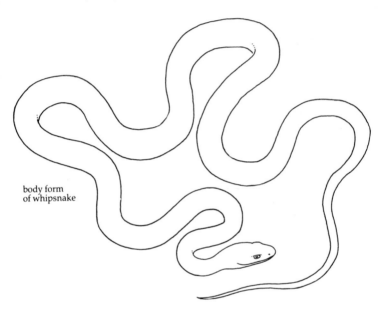

body form
of whipsnake

SONORAN WHIPSNAKE
67½ in. (170 cm) total

STRIPED WHIPSNAKE
72 in. (183 cm) total

PATCHNOSE SNAKES—genus *Salvadora*

Three species, close relatives of whipsnakes, are known in North America. Like most members of the diurnal group, they have a divided anal, 17 rows of smooth scales at midbody, 8–10 upper labials, and two anterior temporals. They have 1–4 loreals and usually 2–3 preoculars and postoculars. Their most distinctive feature is an enlarged rostral, triangular as seen from above, its edges free, giving the appearance of a loosely plastered patch on the tip of the snout. All are striped, fast-moving terrestrial snakes. The ground color is pale gray-brown; they have a broad median yellowish to orangish light stripe that is bordered on each side by a broad black or dark brown dorsolateral stripe, sometimes light-flecked; there is often a narrow lateral dark line. The belly is unmarked and white to yellow, often orange or pink posteriorly. The Mountain Patchnose has two posterior chin shields in contact or separated by one scale; in other species the shields are separated by 2–3 scales. The Big Bend has two scales reaching the eye and one loreal (like the Mountain), but the Western usually has none or only one reaching the eye and rarely fewer than two loreals. Usually there are eight upper labials in the Mountain, 9–10 in the others. In the Mountain, the lateral dark stripe, if visible at all, involves only the second or third scale rows; in others, the fourth row is involved, at least anteriorly.

MOUNTAIN PATCHNOSE SNAKE (*Salvadora grahamiae*). Has 178–197 ventrals, 85–112 caudals. Two subspecies: (1) *S. g. grahamiae*—faint or no lateral line on 3rd scale row; (2) *S. g. lineata*—distinct lateral line.

BIG BEND PATCHNOSE SNAKE (*Salvadora deserticola*). Has 179–205 ventrals, 66–87 caudals. Dorsolateral stripes are distinct, somewhat light-flecked.

WESTERN PATCHNOSE SNAKE (*Salvadora hexalepis*). Has 187–215 ventrals, 73–100 caudals. Three subspecies in U.S.: (1) *S. h. hexalepis*—1 upper labial reaching eye, median stripe 3 scales wide, lateral stripe separate from dorsolateral at least from midbody to rear; (2) *S. h. virgultea*—like (1) but median stripe narrower (1 complete row, 2 half-rows), lateral stripe feebly or not separated from dorsolateral at midbody, top of head brown; (3) *S. h. mojavensis*—usually no upper labials reaching eye, median stripe gray like lateral light areas, dark stripes diffuse, sometimes partially fused as weak crossbars (which may obscure median stripe), top of head gray.

no scales (or only 1) reaching eye and rarely fewer than 2 loreals in Western Patchnose

2 scales reaching eye and 1 loreal in Big Bend Patchnose

1 (or no) scales separating posterior chin shields in Mountain Patchnose

2-3 scales separating posterior chin shields

MOUNTAIN PATCHNOSE SNAKE
47 in. (119 cm) total

BIG BEND PATCHNOSE SNAKE
40 in. (102 cm) total

WESTERN PATCHNOSE SNAKE
47 in. (119 cm) total

195

ELAPIDS—family Elapidae

All snakes with short, fixed front fangs are placed in this family in which several subfamilies are recognized. One is represented in the area covered by this book: Elapinae or Micrurinae (coral snakes). Totally, about 65 genera and 250 species are known in the family, including cobras, kraits, and mambas. Only three genera and 50 species of coral snakes occur in the Western Hemisphere. Their venom is chiefly neurotoxic, adapted primarily for subduing cold-blooded prey. Venoms of some are weak in their effect on humans, but several deaths have occurred from bites of the Eastern Coral Snake. No fatalities are attributed to the Arizona Coral Snake, though it should be regarded as potentially deadly. When startled, coral snakes often hide the head under the body while curling the tail into a ball and waving it from side to side.

The anal and caudals are divided in elapids, which have no loreal, and smooth dorsals in 15 rows for entire length of the body. All coral snakes are oviparous, but some elapids are viviparous. Though essentially nocturnal, all elapids have a round pupil.

Coral snakes in the United States have red, black, and yellow rings encircling the body, the red bordered by the yellow. In similarly ringed but harmless colubrids, the red is bordered by black. A useful ditty: "Red and black, venom lack; red and yellow, kill a fellow." One completely black coral snake has been recorded.

The two species of coral snakes in the United States are distinguished by geographic location and also by differences in arrangement of the anterior rings: in the Eastern, black on the snout is followed by a yellow, then a broad black ring before the typical triads commence; in the Arizona, black on the snout is followed immediately by the first typical triad.

EASTERN CORAL SNAKE (*Micrurus fulvius*). Has 195–232 ventrals; 26–50 caudals; 5–19 (usually 11–15) black rings on head and body, 2–7 (usually 3 or 4) on tail. Yellow rings are relatively narrow; red rings are flecked with black. Habits are as in Arizona. Two subspecies: (1) *M. f. fulvius*—black flecking sparse in red rings, black band on neck not touching tips of parietal scales; (2) *M. f. tenere*—black flecking abundant, black band touching parietal tips.

ARIZONA CORAL SNAKE (*Micruroides euryxanthus*). Has 214–241 ventrals; 21–34 caudals; 9–13 black rings on body, 2 on tail. Yellow (or white) rings are broad. A burrower, it is secretive and nocturnal, active diurnally after rains. One subspecies in U.S.: *M. e. euryxanthus*.

196

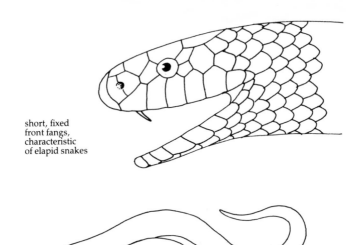

short, fixed front fangs, characteristic of elapid snakes

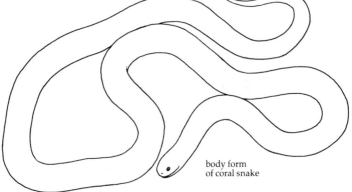

body form of coral snake

EASTERN CORAL SNAKE
47½ in. (121 cm) total

ARIZONA CORAL SNAKE
21 in. (533 mm) total

PIT VIPERS
—family Viperidae, subfamily Crotalinae

This family is comprised of three subfamilies: Viperinae and Azemiopinae (one genus, one species), restricted to the Eastern Hemisphere; and Crotalinae (formerly ranked as a family), confined largely to the Western Hemisphere. The 20 genera and 200 species in the family are split about equally between Viperinae and Crotalinae at the generic level, but about 60 percent of the species are in Crotalinae.

All species of Crotalinae are nocturnal and have vertical pupils. All except the Bushmaster of Central and South America are viviparous, and all have unique facial pit organs (between the eyes and nostrils) that are extremely sensitive heat detectors for finding warm-blooded prey.

Two groups of crotaline snakes occur in the United States: moccasins and rattlesnakes. Moccasins have a pointed tail tip and no rattle; rattlesnakes, found only in the New World, have a very blunt tail tip bearing one or more segments of a rattle.

MOCCASINS—genus *Agkistrodon*

About a dozen species are known, most of them Asiatic. Three occur in the Americas, two in the United States. Both of the latter two have keeled scales, an undivided anal, and undivided caudals (except sometimes near the tip of the tail, which is conspicuously yellow or greenish yellow in young).

The United States species are distinguished by color and habitat. The Cottonmouth is dark and found near water; the Copperhead has a bright pattern of brown crossbands on a tan background and is rarely found near water. (The markings of young Cottonmouths are almost identical to those of the western subspecies of the Copperhead, however.) The Cottonmouth has 23 rows of scales at the midbody, a loreal, a broad dark stripe through the eye, and lips that are often dark; the Copperhead has 25 rows of scales, no loreal, a very narrow or no dark line through the eye, and lips that are never dark.

COPPERHEAD (*Agkistrodon contortrix*). Has 140–157 ventrals, 37–59 caudals. Five subspecies: (1) *A. c. contortrix*—dark crossbands strongly constricted at midline, 2 parts of hourglass shape narrowly connected or evenly separated, often staggered; (2) *A. c. mokasen*—dark crossbands about half as wide at midline as at sides, less often broken, small lateral dark spot in numerous interspaces; (3) *A. c. phaeogaster*—like (2) but interspace spots absent; (4) *A. c. laticinctus*—crossbands scarcely narrower at midline than on sides, belly weakly or not marked; (5) *A. c. pictigaster*—much like (4) but belly strongly pigmented and patterned.

long, hinged
front fangs,
characteristic
of pit vipers

body form
of pit viper

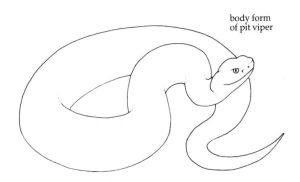

A. c. contortrix

COPPERHEAD
53 in. (135 cm) total

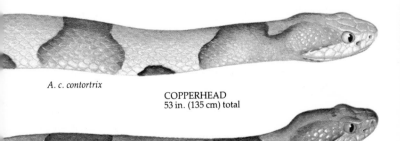

A. c. mokasen

199

COTTONMOUTH (*Agkistrodon piscivorus*). Has 130–144 ventrals, 37–53 caudals. When alarmed, draws head back and snaps mouth wide open, revealing strikingly white interior. Often vibrates tail rapidly (water snakes never do). Three subspecies: (1) *A. p. piscivorus*—evidence of body pattern in all except largest adults, but pattern on sides of head lost early; (2) *A. p. conanti*—like (1) except pattern on sides of head retained throughout life; (3) *A. p. leucostoma*—smaller than (1) or (2), body and facial pattern lost early in life.

RATTLESNAKES

Most viperids can vibrate the tip of the tail rapidly, making a somewhat rattling sound if in dry leaves or some other resonant medium. (Many colubrids may also vibrate the tail.) Only rattlesnakes have a resonator, or rattle, to amplify the vibrations. The rattle presumably evolved, through selection pressure, as a warning device to prevent the snakes from being trampled by the large hoofed animals that were once extremely abundant through much of the Americas. All rattlesnakes have strongly keeled scales, an entire anal scale, and undivided caudals except for a few near the tip of the tail.

The two genera of rattlesnakes—*Sistrurus* (three species, two in the United States) and *Crotalus* (28 species, 13 in the United States)—differ in the arrangement of the scales on the head. Species of *Sistrurus* (plated rattlesnakes) have the nine typical large plates on top of the head; species of *Crotalus* (mailed rattlesnakes) have numerous small scales as well as a few large ones, suggesting the chain mail of antiquity. The Pigmy Rattlesnake, one of the two species of *Sistrurus*, has a remarkably slender tail and a narrow rattle scarcely wider than the eye, whereas in the Massasauga the tail is thicker and the rattle at least twice as wide as the eye. In both species of *Sistrurus* the tail of newborn is bright yellow; it is thought to be used to lure small prey within striking range.

PLATED RATTLESNAKES—genus *Sistrurus*

MASSASAUGA (*Sistrurus catenatus*). Has 21–27 rows of scales at midbody, 129–160 ventrals, 19–34 caudals, 21–50 body blotches. Three subspecies: (1) *S. c. catenatus*—belly mostly black but occasionally heavily blotched, dorsal ground color dark gray to black, body blotches with light outline; (2) *S. c. tergeminus*—belly light with a few dark marks, dorsal ground color light gray to olive-gray, contrasting with dark blotches; (3) *S. c. edwardsi*—like (2) but belly lighter (nearly white), often unmarked, back paler, usually 23 scale rows (25 in others).

A. p. conanti

COTTONMOUTH
74½ in. (189 cm) total

A. p. leucostoma

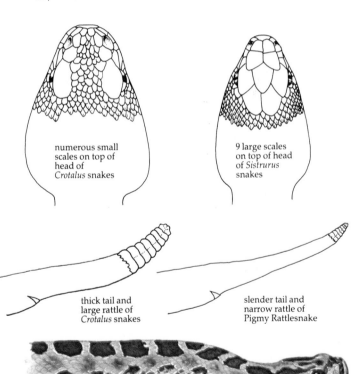

numerous small
scales on top of
head of
Crotalus snakes

9 large scales
on top of head
of *Sistrurus*
snakes

thick tail and
large rattle of
Crotalus snakes

slender tail and
narrow rattle of
Pigmy Rattlesnake

S. c. tergeminus

MASSASAUGA
39½ in. (1,003 mm) total

PIGMY RATTLESNAKE (*Sistrurus miliarius*). Has 122–148 ventrals, 25–39 caudals, 22–45 body blotches. Three subspecies: (1) *S. m. miliarius*—relatively light above, markings sharp, belly light, 1–2 rows of dark lateral spots; (2) *S. m. barbouri*—dark, markings diffuse, belly dark, 3 rows of lateral spots; (3) *S. m. streckeri*—like (1) but 19–23 (usually 21) scale rows, as opposed to 21–25 (usually 23) in others.

MAILED RATTLESNAKES—genus *Crotalus*

Species are described in order—from the more primitive, large, widespread species to the diminutive, specialized high-altitude types.

KEY TO MAILED RATTLESNAKES

1. Scale above eye raised as a short horn . . . . . **Sidewinder,** p. 206
   Not as above . . . . *see* **2**
2. (*not shown*) Crossbands light, dark-edged; no blotches or dark crossbands . **Ridgenose,** p. 206
   Not as above . . . . *see* **3**
3. (*not shown*) 1 row of sharply defined small dark spots on each side of midline . . . .
   . . . .**Twin-spotted,** p. 206
   Not as above . . . . *see* **4**
4. (*not shown*) Vertical light line on each side in front of nostril . . . .
   . **Eastern Diamondback,** below
   Not as above . . . . *see* **5**
5. (*not shown*) Dark crossbands, narrower than interspaces . . *see* **6**
   Blotches, or crossbands wider than interspaces . . . . *see* **7**
6. (*not shown*) Tail light, with rings; W of 99th meridian . . . . .
   . . . .**Rock,** p. 206
   Tail dark, rings scarcely or not evident; E of 99th meridian . . . . . .
   . . . . **Timber,** p. 206
7. Plates above eyes separated by a minimum of 2 scales . . . .
   . . . . **Mojave,** p. 204

   Minimum number of scales between plates more than 2 . *see* **8**
8. Plates above eyes with 1 or more deep grooves; or 2 or more scales between rostral and nostril . . . . . **Speckled,** p. 206
   Neither . . . . *see* **9**
9. 3 or more scales bordering rostral between nasals . . . .
   . . . . **Western,** p. 204
   2 scales . . . .*see* **10**
10. Tail dark, with rings scarcely or not evident, head often black . . . . .
    . . . . **Blacktail,** p. 206
    Tail rings clearly evident. . .*see* **11**
11. Both light and dark tail rings in sharp contrast with body color .
    . . . .*see* **12**
    Tail rings like body colors, small head, wide rattle . . . .
    . . . . **Tiger,** p. 204
12. Reddish, without black peppering in diamonds; SW California . . .
    . . . . **Red Diamond,** p. 204
    Not usually reddish, diamonds black-peppered; SE California and eastward . . . .
    Western Diamondback, p. 204

EASTERN DIAMONDBACK RATTLESNAKE (*Crotalus adamanteus*). Has 25–31 (usually 29) rows of scales, 165–187 ventrals, 20–33 caudals, 24–25 body rhombs surrounded by 1 row of yellow scales. Three or more diagonal light lines on head, rear ones not intersecting mouth. Tail is same color as body, with black rings and black tip.

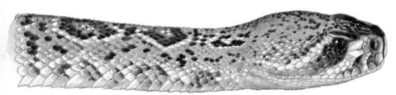

PIGMY RATTLESNAKE
31 in. (787 mm) total

KEY TO MAILED RATTLESNAKES

scale above eye a horn

1

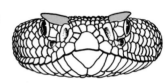

not a horn

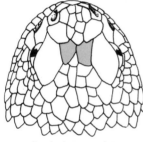

2 scales between plates
above eyes

7

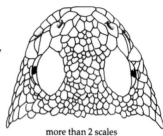

more than 2 scales

8

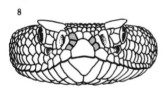

2 or more scales between
rostral and nostril

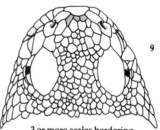

9

3 or more scales bordering
rostral between nasals

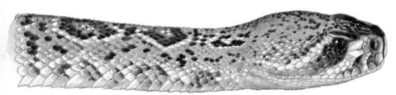

EASTERN DIAMONDBACK RATTLESNAKE
96 in. (244 cm) total

WESTERN DIAMONDBACK RATTLESNAKE (*Crotalus atrox*). Has 23–29 (usually 25) rows of scales, 171–196 ventrals, 16–32 caudals. The 1st lower labial is rarely divided transversely (except in California). Body rhombs number 30–42. Two light lines intersect mouth. Tail is lighter than body and black-ringed.

RED DIAMOND RATTLESNAKE (*Crotalus ruber*). Has 25–31 (usually 29) rows of scales, 185–206 ventrals, 16–29 caudals. The 1st lower labial is divided transversely. Body rhombs, 29–42. Head and tail markings as in Western Diamondback. One subspecies in U.S.: *C. r. ruber*.

MOJAVE RATTLESNAKE (*Crotalus scutulatus*). Has 21–29 (usually 25) rows of scales, 166–192 ventrals, 15–29 caudals. There is a minimum of 2 scales between scales above eye (unique). Rhombs on body number 27–44. Ground color usually greenish. Tail as in Western Diamondback. Line behind eye does not intersect mouth. In proportion to size, most dangerous snake in U.S. One subspecies in U.S.: *C. s. scutulatus*.

WESTERN RATTLESNAKE (*Crotalus viridis*). Has 21–29 (usually 25–27) rows of scales, 161–196 ventrals, 14–31 caudals. Three or more scales border rostral between nasal (unique). Body rhombs number 25–57. Line behind eye does not intersect mouth. Eight subspecies in U.S.: (1) *C. v. viridis*—line behind eye narrow, sharply defined, 13 or more rows of scales at midtail; (2) *C. v. nuntius*—like (1) but 12 or fewer rows of scales at midtail, reddish, adults rarely over 25.5 in. (650 mm); (3) *C. v. concolor*—line wider, diffuse, or lacking, yellowish, adults less than 25.5 in. (650 mm); (4) *C. v. abyssus*—like (3) but vermilion or salmon, blotches indistinct; (5) *C. v. lutosus*—like (3) but not yellow, larger, spaces between blotches on rear of body equal to or wider than interspaces; (6) *C. v. cerberus*—like (5) but spaces narrow, scale in front of nostril touching upper labials; (7) *C. v. helleri*—like (6) but blotches longer, scale touching upper labials, last tail ring twice as wide as others and ill-defined; (8) *C. v. oreganus*—like (7) but tail rings uniform, clearly defined.

TIGER RATTLESNAKE (*Crotalus tigris*). Has 21–27 (usually 23) rows of scales, 158–177 ventrals, 16–27 caudals, and 37–52 body blotches in the form of dark crossbands. The head is small; the rattle is large.

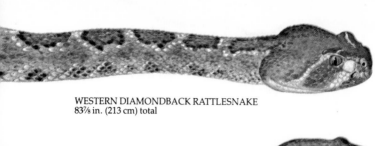

WESTERN DIAMONDBACK RATTLESNAKE
83⅞ in. (213 cm) total

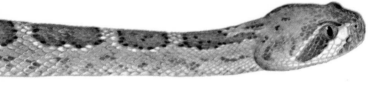

RED DIAMOND RATTLESNAKE
65 in. (165 cm) total

MOJAVE RATTLESNAKE
51 in. (130 cm) total

C. v. viridis

WESTERN RATTLESNAKE
64 in. (163 cm) total

C. v. oreganus

TIGER RATTLESNAKE
36 in. (912 mm) total

SPECKLED RATTLESNAKE (*Crotalus mitchelli*). Has 21–27 scale rows, 166–187 ventrals, 16–28 caudals, 23–43 body blotches. Two subspecies in U.S.: (1) *C. m. pyrrhus*—nostril separated from rostral by 2 or more scales, narrow tail rings; (2) *C. m. stephensi*—1 scale between nostril and rostral, supraoculars partly divided, tail rings wider.

SIDEWINDER (*Crotalus cerastes*). Has 19–25 (usually 21–23) rows of scales, 132–154 ventrals, 14–26 caudals, 28–45 body blotches. Elevated, conical scales above eye. Sidewinding dweller of sandy wastes. Three subspecies: (1) *C. c. cerastes*—basal rattle segment brown; (2) *C. c. cercobombus*—basal segment black, usually 21 scale rows, 141 or fewer ventrals in males, 145 or fewer in females; (3) *C. c. laterorepens*—like (2) but 23 scale rows, more ventrals.

BLACKTAIL RATTLESNAKE (*Crotalus molossus*). Has 25–29 (usually 27) rows of scales, 178–199 ventrals, 18–29 caudals, 22–39 body blotches. One subspecies in U.S.: *C. m. molossus*.

TIMBER RATTLESNAKE (*Crotalus horridus*). Has 21–26 (usually 23–25) rows of scales, 158–183 ventrals, 13–30 caudals, 15–34 body crossbands. The lowland *C. h. atricaudatus* (formerly recognized) has a dark band which runs down diagonally behind the eye.

RIDGENOSE RATTLESNAKE (*Crotalus willardi*). Has 25–27 rows of scales, 147–159 ventrals, 21–35 caudals. Upper edge of snout is ridged; 19–27 crossbands. Two subspecies in U.S.: (1) *C. w. willardi*—vertical light line at tip of snout; (2) *C. w. obscurus*—no line.

ROCK RATTLESNAKE (*Crotalus lepidus*). Has 21–25 (usually 23) rows of scales, 150–172 ventrals, 16–29 caudals, 13–24 bands. Mountain rock-dweller. Two subspecies in U.S.: (1) *C. l. lepidus*—mottled between bands, dark facial line; (2) *C. l. klauberi*—little or no mottling, facial line dim or absent.

TWIN-SPOTTED RATTLESNAKE (*Crotalus pricei*). Has 21–23 (usually 21) rows of scales, 149–171 ventrals, 18–30 caudals. Scale behind nostril contacts upper labial; only 1 complete scale row between eye and labials (unique to this and Rock). Blotches, 41–61. Mountain rock-dweller. One subspecies in U.S.: *C. p. pricei*.

SPECKLED RATTLESNAKE
52 in. (132 cm) total

C. m. pyrrhus

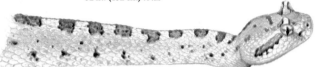

SIDEWINDER
31 in. (787 mm) total

C. c. laterorepens

BLACKTAIL RATTLESNAKE
32½ in. (826 mm) total

TIMBER RATTLESNAKE
74½ in. (189 cm) total

RIDGENOSE RATTLESNAKE
25⅛ in. (635 mm) total

C. l. klauberi

ROCK RATTLESNAKE
32⅝ in. (828 mm) total

TWIN-SPOTTED RATTLESNAKE
26 in. (660 mm) total

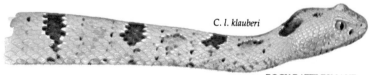

CROCODILIANS—order Crocodylia

This is an ancient order, with 13 families and many genera and species known only as fossils. It is represented today by 21 species in a single family, Crocodylidae, which is worldwide in distribution in the tropics and subtropics. *Crocodylus* contains 11 species, mostly in the Eastern Hemisphere; *Caiman*, two species in Central and South America; *Alligator*, two species—one in North America, one in China. In North America, the crocodile belongs to the subfamily Crocodylinae; the alligator and the introduced caiman to Alligatorinae. A third subfamily, Gavialinae, is limited to India.

All crocodilians are semiaquatic, living in swamps, ponds, lakes, and marine estuaries. They frequently sun on land near water and build nests on land; all lay eggs. All are carnivorous. Crocodilians are distinguished from other reptiles by a longitudinal anus (transverse in snakes, lizards, and amphisbaenids; round in turtles). Males have a single copulatory organ, as do turtles (tuatara have none; snakes, lizards, and amphisbaenids have two). Valved nostrils and ear openings can be closed underwater. Nostrils and eyes are protuberant, enabling the animal to see and breathe without exposing the rest of its body. A valvular tongue permits closure of the passage to the lungs while food is macerated partially underwater. Ventral scales are large, quadrangular, and soft; the larger dorsal scales are over bony plates; lateral scales are small and soft; a dorsolateral row of protuberant scales at each side of base of tail joins as one row at about the middle.

AMERICAN CROCODILE (*Crocodylus acutus*). Has enlarged 4th tooth of lower jaw conspicuous at sides of upper jaw when mouth is closed, as are most other lower jaw teeth. Tip of snout is somewhat widened; sides of head strongly taper in dorsal profile. Light gray in general color. One subspecies in U.S.: *C. a. acutus*.

AMERICAN ALLIGATOR (*Alligator mississippiensis*). Has enlarged 4th tooth of lower jaw not protuberant when mouth is closed (enters socket in upper jaw, as do other lower jaw teeth). Snout is broad, tapering little. Dark in general color. Young have yellowish crossbands.

SPECTACLED CAIMAN (*Caiman crocodilus*). Has moderately tapered snout, tip not expanded. Enlarged 4th tooth of lower jaw enters socket in upper jaw, as do other lower jaw teeth. Has conspicuous curved ridge between anterior edges of orbits. Released in the U.S., but no breeding colonies reported. [No range map.]

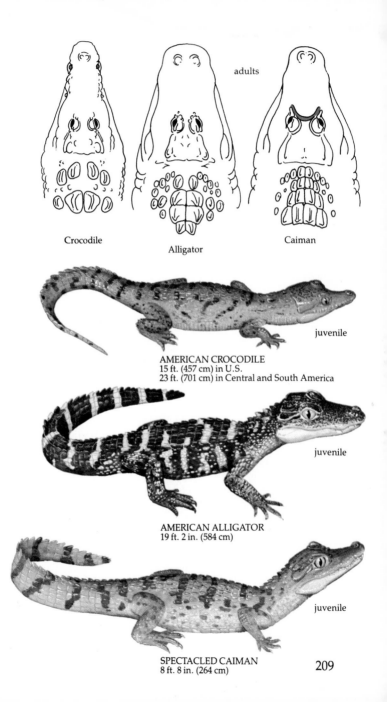

adults

Crocodile

Alligator

Caiman

juvenile

AMERICAN CROCODILE
15 ft. (457 cm) in U.S.
23 ft. (701 cm) in Central and South America

juvenile

AMERICAN ALLIGATOR
19 ft. 2 in. (584 cm)

juvenile

SPECTACLED CAIMAN
8 ft. 8 in. (264 cm)

209

REPTILES

Evolution

Reptiles were the first vertebrates to escape dependency on water, a restriction that confines the amphibians to at least moist surroundings.

Reptiles first appeared about 300 million years ago in Pennsylvanian times, with the earliest forms showing a mixture of amphibian and reptilian characteristics. The reptile group diversified greatly over the next 200 million years. During the Mesozoic, they were the dominant animals on earth. At this peak period, reptiles were represented by some 15 major groups. Only four of these orders survive today.

Extinct are the fishlike ichthyosaurs, sail-backed pelycosaurs, flying pterosaurs, mosasaurs, plesiosaurs, well-known dinosaurs, and many others. The dinosaurs, or "terrible lizards," included the largest animals ever to walk on earth—the sauropods, some of them nearly 90 feet (27 m) long. Many of the less familiar dinosaurs were no larger than chickens.

Several basic advances made possible the rise and wide distribution of reptiles on land. Most important were the amniote egg, with its tough outer covering and protective membranes, and a scaly or cornified skin that protected the animals from drying out. The positioning of the limbs also made it possible for reptiles to move more easily on land, and the improved circulatory system (an incompletely five-chambered heart) gave them a more oxygen-rich blood supply.

In their Mesozoic heyday, reptiles dominated the land, seas, and air. The reason for their rapid decline at the end of the Mesozoic is still not understood, but at this same time the warmblooded vertebrates—birds and mammals—began their expansion. By Cenozoic times, only four orders of reptiles still existed, and these same four have persisted to this day. The order Rhynchocephalia is represented by only one species, the lizardlike, granular-scaled Tuatara (*Sphenodon punctatus*) confined to New Zealand where its survival is now threatened. The remaining three orders have representatives throughout the world.

The order Testudines (turtles) is the most ancient, appearing about 250 million years ago and remaining virtually unchanged for the past 200 million years. The order Crocodylia (crocodilians) is only slightly less ancient and is traceable to the Permian thecodonts. The order Squamata (scaled reptiles—lizards, amphisbaenids, and snakes) is the most recent, not common until late Cretaceous times only about 65 million years ago.

Distribution

Modern reptiles are only remnants of the great numbers and variety that existed in the past, and they occur most abundantly, both as species and as individuals, in the subtropics and the tropics. Reptiles become increasingly scarce toward the poles and also at high altitudes, temperature being the primary limiting factor. In North America, reptiles are moderately abundant in the southern third, much less common in the middle third, and scarce to absent in the northern third.

Scales and Scutes

Development of a dry skin was one of the major advances of reptiles over amphibians, whose moist and glandular skin made them dependent on aquatic or semiaquatic surroundings. In scaled reptiles (lizards, amphisbaenids, and snakes), the outer and cornified (horny) epidermis (ancestrally covering the dermal scales below) is shed periodically, less frequently by lizards than by snakes, which in protracted warm weather and with abundant food may shed as many as half a dozen times a year. The average frequency is three times per year, but some molt only once a year. At shedding time, the skin begins to loosen over a snake's body. The caps over its eyes, consisting of a single transparent scale, first become "cloudy" or "milky" due to moisture from cell degeneration between the old and the new layers. A few days later the eyes clear, and the snake finds a rock, limb, or similar object on which it begins to rub its nose. This loosens the skin, and the snake crawls out, commonly leaving the "inside out" shed skin entire.

Some of the smaller lizards also molt in this manner, but lizards most commonly shed their skin in patches. Geckos and some other lizards eat their freshly shed skin.

The scales of snakes and many lizards overlap with soft, flexible skin between cornified regions. This allows the bodies of these animals to expand with a large meal.

The horny scutes of a turtle's shell are also covered by an epidermal layer that is shed at irregular intervals by some species. The leathery skin of crocodilians covers deep bony dermal plates. The skin is not shed or molted but is replaced by growth from beneath as it is worn away.

Chromatophores, or pigment cells, in the skin give reptiles their colors and patterns. These include melanophores (black or brown) and lipophores (red, orange, and yellow). In addition, many reptiles have guanophores bearing a white substance and iridophores that lack pigments but produce iridescent colors (green, blue) prismatically by light refraction. Anoles can change their color (metachrosis) from green to brown and to intermediate combinations. Color changes are apparently induced by hormones in the reptile, the glands functioning in response to a variety of environmental disturbances.

Skeleton

Compared with the amphibians from which they were derived, reptiles have a much greater amount of bone in their skeletons and less cartilage.

A snake's jaws are unique, enabling it to swallow prey much larger in circumference than the snake's normal mouth opening. Features making this possible are the joining of the two halves of the lower jaw by a stretchable ligament so they can be spread apart, the loose junction of the lower jaws to the upper at the rear so the mouth can be opened wide as prey is moving in, and the loosely joined bones in the roof of the mouth not forming a rigid palate and thus allowing them to spread apart.

In lizards and crocodiles, the lower jaws are joined by a suture, hence there is no stretchable ligament. In turtles, the bones are fused to form a single, solid mandible. In crocodilians, a flaplike velum at the back of the mouth cavity shuts off the connection between the mouth and the air passages for the nostrils so the animals can drown their victims by holding them in their mouth underwater and can continue to breathe themselves as long as their bulbous nostrils are above the surface. This is made possible by a fusion of the bones in the roof of the mouth to form a complete hard palate that extends far back into the throat.

Snakes are distinguished by their lack of legs, and a few lizards are also limbless. But the snakes have also lost their pectoral and pelvic girdles. Vestiges are evident only in primitive groups, such as blind snakes, boas, and pythons. Most other reptiles have two pairs of limbs (a few have a single pair), typically with five toes that end in horny claws.

Turtles are unique in having both pectoral and pelvic girdles inside their rib cage, a transformation that occurs during their embryonic development.

As with other vertebrates, the total number of vertebrae, or back-bones, remains the same throughout the animal's postembryonic life, except for a possible reduction due to fusion as an animal gets older. Among vertebrates, snakes have the greatest number of vertebrae. Pythons have the most—to more than 400—but in native North American snakes, the number of vertebrae is commonly 250 or more. This gives the sinuous snakes their remarkable flexibility. In contrast, a giant alligator has only 26 vertebrae.

A turtle's legs project at the sides of the body, essential because of the shell encasement. In crocodilians, too, the legs are spread, but an alligator can nevertheless lift its heavy body on its legs and run incredibly fast for short distances. In lizards, the legs are more nearly under the body. Many lizards are fleet, and the fastest are bipedal—running on their hind legs only. Many lizards are also adept at climbing.

The locomotion of the legless snakes is unique. Snakes are sinewy because of the many muscles in their body, connecting the vertebrae to each other and also to the skin. This complex musculature makes possible the snake's gracefulness. In fast crawling, a snake extends small lateral loops (kinks) of its body and engages them on irregularities in the surface over which it is moving. "Sidewinder" rattlesnakes have exaggerated this looping so they move at a 45-degree angle to the position of their head and tail.

A snake appears to move much more rapidly than it actually does. Three miles (4.8 kilometers) an hour is fast for most snakes, though it would be difficult to convince most people that the speed is not much greater. When it is moving slowly and in a straight line, a snake employs the broad plates on its belly, engaging them in any irregularities on the surface. Climbing snakes also make great use of their ventral (belly) scales in addition to kinking their body to maintain their hold on branches.

Digestive System

The digestive system of reptiles is essentially like that of other vertebrates with only a few notable specializations. Reptiles have lost the numerous mucous glands characterizing amphibians but are still well endowed with salivary glands. Reptiles swallow their food whole or in big chunks, and moistening the food with secretions from the glands in the mouth aids in the process. Because they are mainly carnivorous, reptiles have strong digestive enzymes. This makes possible digesting their prey whole without chewing.

The poisons of venomous snakes are produced by enlarged and modified labial (Duvernoy's) glands. The venom of the poisonous lizards (Gila Monster and Mexican Beaded Lizard) is produced by modified sublingual glands located along the lower jaw.

Crocodiles and turtles have broad, flat tongues that are short and cannot be protruded from the mouth. In Alligator Snapping Turtles, a wormlike forked extension at the rear of the tongue can be distended and used as a lure for attracting small fish or other prey. In both lizards and snakes, the tongue is long and extensible, less so in North American lizards than in the African chameleons and some others. The geckos have a remarkable tongue—long enough so they can clean their transparent eyelids with it.

The forked tongue of snakes and some lizards is an unusual sensory organ that can be protruded through a notch in the middle of the lower jaw without opening the mouth. The snake "flicks" its tongue in and out to sense its surroundings chemically. Each time the tongue goes back into the mouth, the forked tips are inserted into grooves in the roof of the mouth. The grooves connect to Jacobson's organ in the nasal cavity. In this way—by chemical sensing—the snake smells its surroundings.

Respiratory System

Reptiles have lungs and are air breathers. Even giant sea turtles, the most aquatic of the reptiles, must surface from time to time to take in air. As part of their elongation, nearly all snakes (some primitive types excluded) have only one lung, usually the right. Some lizards can inflate their lungs to disproportionate size. Chuckwallas use this device to expand their body and wedge themselves into crevices. Marine turtles inflate their lungs to help keep themselves afloat when they are basking at the surface. Some aquatic freshwater turtles, particularly the softshell turtles, and possibly also the marine turtles, may achieve some exchange of gases through highly vascularized areas of their skin and probably also through the cloaca.

Circulatory System

Amphibians have a single ventricle so that a mixture of aerated and non-aerated blood is pumped through the body. Reptiles have a septum in the ventricle—incomplete in most but complete in the crocodilians. But even the partial septum reduces the mixing of the oxygen-enriched blood from the lungs with the oxygen-depleted blood from the body, thus resulting in greater efficiency. Amphibians have a separate chamber, the conus arteriosus, in the heart. In reptiles, this chamber is transformed into the bases of the great arteries—pulmonary and aortic—conducting blood from the heart. Two separate atria and a small sinus venosus are retained—thus reptiles possess five chambers of the heart.

Excretory and Reproductive Systems

Lizards and alligators have elongate kidneys much like those of amphibians. Turtles have quite compact kidneys. In snakes, the two kidneys are stretched out and located one behind the other, or in tandem, rather than side by side. Turtles and some lizards have a bladder, but snakes, crocodilians, and most lizards lack a bladder. In those with a bladder, excretory wastes from the kidneys are first stored there and then eliminated via the cloaca, the common chamber that receives wastes from the digestive tract, excretions from the kidneys, and also reproductive products. The name cloaca, in fact, is derived from the Latin word meaning sewer.

Fertilization is internal in all reptiles. Except in *Sphenodon*, males are equipped with copulatory organs—double in lizards, amphisbaenids, and snakes (hemipenes), single in turtles and crocodilians. The single organ lies along the posterior wall of the cloaca and is protruded during mating. The hemipenes lie at the base of the tail but open, one on each side, into the rear cloacal wall. In mating, one or the other is everted. Sperm move along a groove in these organs and into the female's cloaca, fertilization occurring in the oviducts. After mating, the oviducts of some snakes are sealed with a secretion, preventing further copulation until eggs are laid or young are born.

Most reptiles lay eggs (oviparous). Some give birth to their young (viviparous). The viviparous species (garter snakes and night lizards, for example) in which the embryos lack a shell but have placental modifications facilitating exchanges with the mother are euviviparous, much like most mammals. All viviparous reptiles that have been examined in adequate detail are euviviparous, but it is possible that some lack placental specializations (like papillae) for maternal-embryonic exchanges. If so, they may be regarded as ovoviviparous.

Nervous and Sensory System

An advance over amphibians, reptiles have 12 cranial nerves as opposed to 10. The brain is still rather primitive, but the locomotor center, the cerebellum, is well developed whereas it is very small in amphibians. In some of the extinct dinosaurs, in fact, the pelvic enlargement of the spinal cord controlling motor functions in the hind legs and the rear portion of the body was actually larger than the creature's brain. Many amphibians, particularly the more aquatic ones, have a pressure-sensitive lateral line system, but this is completely lost in reptiles. In other respects, reptiles are more advanced in sensory and motor responses than are amphibians but far less so than birds and mammals.

All reptiles respond to vibrations, though only lizards and crocodilians have external ear openings. Snakes are functionally deaf to airborne sounds, even though their auditory apparatus reacts to sound waves much as in mammals. They are highly sensitive to vibrations in the substrate. Cobras rise in response to the rhythmic swaying of the snake charmer's body rather than to the sound of music.

Some snakes, most notably the pit vipers and some members of the boa family, have unusually well-developed thermoreceptors. These heat-sensing organs are located in the loreal pit on each side between the nostril and the eye in rattlesnakes and other pit vipers and along the lips in boas and pythons. When hunting, these reptiles do not need to search over the total area of a branch or the complete labyrinth of a burrow, for they can sense remotely whether a potential meal is present by the heat given off from its body.

Almost without exception, reptiles have lateral vision. They cannot focus both eyes on the same object directly ahead. Snakes, geckos, and some limbless lizards do not have movable eyelids. This gives them the unblinking stare that has resulted in the folk belief that snakes can hypnotize their prey. A transparent scale actually covers each eye, protecting the eye from sand, twigs, or other objects. In most reptiles, the pupil is round, but in some, such as the vipers and the nocturnal lizards, the pupil is slitlike.

REPRODUCTION

As among most animals, mating is a spring ritual for the majority of reptiles. The booming roars of male alligators in early spring announce their breeding season, but in other reptiles, there are no noisy congresses of males as in amphibians. With only a few exceptions, mating is quiet and unnoticed.

Some snakes court before mating, the male rubbing his chin over the female's body and the two sometimes gliding off side by side or entwining their bodies. Courting may continue for an hour or more. After a single mating, a female snake may continue to produce young or to lay fertile eggs for two to seven years or longer.

Many male lizards perform elaborate courtship displays. Males of some species of lizards have glands in scales on the rear venter or the undersides of their hind legs. The secretions from these glands harden, forming protrusions somewhat like the teeth of a comb. They are used in courtship and in spreading pheromones (chemical attractants), and they give the male a more secure grip on the female during mating.

The mating of turtles is quiet and unostentatious, with usually brief courtship performances, brutally physical in some species, touchingly tender in others. Turtles need a lot of energy to accomplish copulation, an especially tricky feat for the land dwellers and only somewhat less so for aquatic species.

Eggs

Female sea turtles may lay as many as 400 eggs in a season. The eggs are deposited in a scooped-out hole just above the high-tide mark on the beach. The annual trip of the female to shore, laboring ponderously over the sand, is one of nature's wonders and is the only time that marine turtles leave the sea. Each female may actually make several trips before a suitable place to dig is found, and some females make several nests every season, depositing eggs in each.

Because sea turtles have been overharvested for food and many of their nesting sites have been destroyed by encroaching civilization, all sea turtles are now threatened or endangered species. In the United States, their egg-laying forays are protected events. Dogs, raccoons, and other animals in addition to humans must be kept away from the eggs, too, which have long appealed to predators.

Other turtles lay far fewer eggs, rarely more than a dozen. Aquatic turtles go through the same process as sea turtles, scooping out a hollow in the sand or in debris at the water's edge. Land turtles lay their eggs in litter and loose debris. The heat of the sun plus the heat generated by the decaying debris hatches the eggs, for as soon as the eggs are laid, the female departs and shows no more interest.

Crocodilians commonly lay several dozen eggs—sometimes to as many as 60. Some bury their eggs rather casually in dirt or moist debris, but the alligator constructs an elaborate nest, the female pushing together heaps of rotting debris and then laying her eggs in a depression in the middle of this mound. More vegetation and debris are heaped on top, and the heat from the decomposition plus the heat of the sun hatches the eggs. The female does not leave her nest. She stays nearby and defends the mound from marauders. She may also add more moist vegetation from time to time.

At hatching time, the baby alligator in each egg begins to squirm. On top of its snout, the hatchling has a tiny, horny conical "egg tooth" used to break through the shell of the egg. Suddenly two dozen or more baby alligators are squealing and grunting inside the heaped pile, and the mother alligator hears them. She grunts in reply, then climbs onto the heap and begins to tear it apart. This helps the babies to escape. The female carries her young, each 6 to 8 inches (152–203 mm) long, to the water in her mouth, one or a few at a time. The young apparently stay with their mother through the first winter. She gives them no help in getting food but may respond to distress cries and protect the babies from predators.

Baby lizards and snakes also have an "egg tooth" at the tip of their snout. The egg tooth is shed within a few days, having served its function of slicing through the egg's leathery skin. Lizards and snakes lay a few to as many as two dozen eggs. Like alligator eggs, they are long and slim, as big at one end as at the other, but they usually have a tough outer covering rather than a limy shell. Snakes and lizards do not, with exceptions, stay with their eggs or give their young attention.

Numerous species of snakes and lizards are live-bearers (viviparous). The eggs retained in the female's body never develop a shell, and in some an elaborate exchange system exists between the embryo and the mother. This exchange structure (placenta) is less complex than in most mammals but is fundamentally the same.

Growth and Longevity

At hatching or birth, young reptiles have the same body form as their parents and are equipped to fend for themselves from the start. Young poisonous snakes, for example, have functional venom glands. The young of some reptiles may have a different color or pattern from their parents, and such differences are noted in the text in this book. Young reptiles grow very rapidly during the first year or two, most becoming sexually mature in their second or third years. Alligators are not ready to mate until they are five or six years old.

The longevity of reptiles, particularly turtles, is legendary. Some records are disputable, but there is substantial evidence that a number of species of turtles do live to be 75 to 100 years old. Crocodilians, lizards, and snakes are probably ancient in nature at age 20 but have survived for 40 years or even longer in captivity where they get good attention and are protected from predators and younger, more vigorous competitors. At the opposite extreme, many small species, particularly lizards, have a life expectancy of only a year or even less (11 months in the Green Anole, for example).

DORMANCY

Reptiles (ectotherms, as are fishes and amphibians) are active during all warm months, but in winter they go into a state of physiological dormancy called brumation, as distinguished from the hibernation of birds and mammals (endotherms). This period of inactivity may be brief in southern United States, but in northern states, in Canada, and at high altitudes in mountainous regions, the dormancy may last for months. During this time, their metabolic rate is reduced to a very low level. Reptiles prepare for brumation by entering burrows, crawling deep into crevices, or finding other protected places. Rattlesnakes and many other snakes may congregate in the same den. Freshwater turtles drop to the bottom of ponds, lakes, or streams and bury themselves there in mud and debris. While dormant, they respire by the exchange of gases through their skin and cloaca, but their needs are small.

Although reptiles are most active in warm weather, they cannot tolerate excessive heat. In deserts, they avoid the peak heat of the day by remaining in hiding. Reptiles will die more quickly from overheating than from cold. Most reptiles inhabiting the deserts of the southwestern United States, for example, are most active when the temperature is about 100 degrees F. (37 degrees C.), but they cannot survive a body temperature exceeding 112 degrees F. (44 degrees C.).

ENEMIES AND SELF DEFENSE

Compared to amphibians, reptiles are much better equipped to protect themselves. In danger, box turtles hide inside their own shell, and for normal predators (excluding humans), this is sufficient. Only when the outside world becomes quiet again does the turtle venture first a peek through a slitlike opening of its shell and then, if not startled, open its shell completely to go plodding on.

Aquatic turtles, lizards, snakes, and crocodilians—all first try to escape danger by running, hiding, or keeping very quiet. Generally they escape detection. If cornered or caught, these passive tactics may change to aggressiveness as a matter of survival. Alligators lash their massive tail, or if sufficiently agitated, an alligator may charge an intruder. When an alligator gets a firm grip with its tooth-studded jaws, it then rolls its body. This is the way it suffocates large prey underwater or tears chunks from its body.

Toothless turtles can give severe bites, the sharp edges of their jaws cutting deeply. Some turtles cannot be induced to bite, but others—notably snappers and softshell turtles—tilt toward an intruder ready to strike. They can lash out and twist their snakelike necks to score hits at astonishing distances and directions. Lizards can inflict painful but harmless bites, the exceptions being the venomous Gila Monster and Mexican Beaded Lizard. Most snakes in North America are also harmless, but all except the diminutive blind snakes have numerous backward-curved teeth. When a snake strikes and then draws back, these relatively small but hooked teeth usually tear the skin and cause profuse bleeding. Often one or more teeth break off in the wound, which should be treated with an antiseptic.

Many lizards shed their tail (autotomize) when it is grasped by a predator. The predator consumes the tail while the lizard escapes. The tail is shed without bleeding, and exposed nerve endings in the tail cause it to wriggle for several minutes. The tail is regenerated to nearly its original size, but the vertebral column of the original tail is replaced by a cartilaginous rod in the regenerated tail. Geckos and some other lizards posture with their tail elevated and waved in the direction of the predator. The attacker is drawn to the tail, which is readily shed. Some lizards and many snakes cannot regenerate their tail if it is lost, and no reptiles can regenerate limbs.

Horned lizards are protected from predators by the crown of spines on their head. When attacked, they turn these spines toward the enemy. They also startle attackers by squirting blood from the corners of their eyes.

Some snakes, like the Rubber Boa, coil their body with their head hidden under the coils and their tail lifted, appearing like the head. Other harmless snakes hiss and flatten their head; some, such as the bull snakes, put on a display more fearsome than the dangerous rattlesnakes. Hognosed snakes first try a fake attack. If that fails, they turn onto their back, loll their mouth open, and void their cloaca. This behavior is lost quickly in captivity.

In their natural world, all reptiles have a share of enemies. The young are particularly vulnerable to predation. Young sea turtles, for example, must wobble their way feebly from their sand nest to the sea, often a considerable distance and especially hazardous when being able to hurry would help them escape watchful gulls, raccoons, and other predators that are ready to gobble them. Even when they do reach the water, their troubles are not over, for while they are still fresh from the egg, the little turtles are carrying a "lunch box" of nutrition from their embryonic enclosure. This remaining food supply causes the little turtles to bob about on the surface like corks at first. Several hours pass before the turtles are able to submerge to escape predators that are anxious to pick them up at the surface.

Baby crocodilians, lizards, and snakes are also relatively easy prey for birds, turtles, big frogs or toads, and other creatures. This hazard continues for many species on into their adult lives, for small snakes and lizards are the constant prey of larger animals. Even some kinds of lizards and snakes prey on still smaller reptiles.

The reproductive potential of reptiles is, however, balanced to withstand natural enemies and hazards. Population voids that are the result of usual predation are filled in quickly by nearby surpluses, and a reasonable stability of numbers is maintained. As with all animal populations, each species occupies available areas of optimum habitat, expanding at the periphery to the limits of inhabitable area.

The Human Factor

All of the natural balances and counterbalances are put asunder by humans—the greatest enemies of reptiles. Humans devastate reptiles in two ways: by habitat destruction and by the slaughter of individual animals.

The multitudinous ways by which humans "advance" themselves are all destructive to the natural world and its creatures. This generally applies not to reptiles alone, though in some circumstances reptiles may be especially hard hit. The clearing, scouring, and cultivating of lands for agriculture; the draining of wetlands; polluting of ponds, lakes, and streams; building on beaches to the water's edge; the spraying of chemical poisons to kill insect pests and the inadvertent killing of reptiles as a result—these and other alterations of the natural environment have taken a large toll of reptiles.

Highways crisscrossing the land have also brought on a great hazard to natural populations. Turtles, for example, cannot cross a highway with enough speed to escape the fast-rolling wheels of trucks and automobiles. On cool nights, snakes make the mistake of stretching themselves on the still-warm hard surface and thus become highway victims. But these habitat ills that beset reptiles are shared with other animals and also with many plants. Reptiles have still more hazards that are peculiar to them as a group or to only particular species.

Reptiles are the favorite animals of few people. Some people, in fact, seem bent on exterminating reptiles, particularly snakes, and will go out of their way to kill them. Snakes are, in fact, probably the most maligned of all creatures on earth. Few large snakes can escape slaughter when they are encountered by humans. Smaller species are spared only because of their secretive habits and their adeptness at escape. In North America, most snakes are not only harmless but also beneficial because of their diets of insects and rodents that are pests to crops and stored products. The killing of poisonous snakes might be justified if the snakes frequent or have intruded on habitations, but if it is the humans who have invaded the snakes' shrunken world, the killing is rarely necessary. It is better, if it can be done safely, to transport the snakes to nearby wildlife sanctuaries where they can find their niches and live unobtrusively.

A number of species of turtles have suffered near annihilation because they are good to eat. Diamondback Terrapins, which inhabit the brackish waters of the Southeast, became a fad among gourmets a few years ago, and they were harvested to the brink of extinction. Fortunately for the terrapins, the fad faded. Laws were also passed to protect the terrapins. Some are still being produced on "farms" where breeding stocks are also maintained.

Giant sea turtles, particularly the Green and the Loggerhead, met a similar fate, the turtle industry in Florida literally harvesting itself out of business. International agreements have now been established to protect these gargantuans of the chelonian clan, and thousands of baby turtles are released from hatcheries every year in an effort to reestablish the populations of the turtles in the wild.

Alligators are a classic and well-known story in North America. Once abundant throughout Florida and the lower South, alligators came close to extinction because of destruction of their habitat combined with continued intensive killing by hunters and the collecting of young 'gators for sale as pets. In this case, as proved by later happenings, hunting was unquestionably the principal factor in the decline of the alligator. By the hundreds of thousands, alligator hides were converted into leather to make shoes, bags, wallets, belts, and similar items. Scattered populations persisted over their range, but the last major stronghold for the alligator was in the Everglades at the extreme southern tip of Florida. Part of this habitat, specifically the Cape Sable region, was shared with the few remaining American crocodiles, which were never as numerous or widespread but which suffered a similar fate.

With the passing and enforcement of protective laws, most particularly the Endangered Species Act of 1973, the alligator began to make a comeback. It has done so remarkably well that people have begun to complain about the giant beasts as pests and hazards, not only in the Everglades region but also in other areas throughout the state. Alligators are no longer listed as endangered. Controls still exist, but they have been relaxed considerably to allow limited harvesting outside parks, refuges, and other sanctuaries.

As with other creatures, the species of reptiles with the most limited ranges must be watched most closely in terms of their survival. A single dam, a gigantic development, or other environmental modification could eliminate them. Conditions of this sort are being examined and monitored more carefully today as concern about the natural world becomes greater. That concern has led to the passage of not only Federal but also International and State laws severely limiting the legal acquisition of most species of reptiles and amphibians. Enforcement can be extremely stringent. However, the greatest hazard—habitat destruction—is poorly controlled.

STUDYING REPTILES

Reptiles can be studied in nature as they go about their normal activities, in captivity by observing those activities that more or less approximate their behavior in nature, and by experimentation both in nature and in the laboratory.

1. Nature. The most useful observations are those made in nature. Very little is known about the daily habits of most reptiles, even the most common species. People tend to ignore reptiles, or they kill or collect them. Few observers have taken the time to watch them going about their daily lives, observing how they feed, court, mate, lay eggs or give birth to their young. Little is known about how reptiles interact with members of the same or different species, what their reactions are to enemies or how enemies are avoided. Careful studies of this sort are much needed and will become significant contributions to knowledge.

For intensive studies of observable populations, the animals can be marked in various ways to identify them as individuals. Dots of different colors or a water-repellent paint can be used. In the case of snakes and lizards, these will last only until the animals shed. For turtles and crocodilians, the marks will last much longer. Plastic tags, either bands or buttons, are also useful in some studies, but they must be used with care because they may hamper the animals.

Alligators and the larger turtles, lizards, and snakes are now tracked by equipping them with small transistorized transmitters that give off identifying "beeps," some of the signals on larger transmitters detectable to distances of eight or more miles when the animals are on the surface. The distance the transmissions can be picked up when the animals go underwater or underground is much less, of course. This technique, called biotelemetry, has proved very useful in determining where and how fast these reptiles travel and what they do in their wanderings. Most of the studies so far have been concentrated on large animals that have a significant economic importance over wide areas, such as the alligator and sea turtles. But similar studies are needed for smaller reptiles.

2. Captivity. Observing animals in captivity can be helpful, especially when the same animals are also studied in nature. The situations can only be comparable, however, not identical, and it must be clearly understood that the animals might react differently because of their confinement.

3. Experimentation. A variety of experimental studies can be made with captive animals in the laboratory without killing them. Other experiments, such as homing and orientation, can be done in nature with captive animals. When the studies are completed, the animals should be released in their natural habitat, preferably in the exact area where they were collected.

Some animals are sacrificed in professional laboratory studies to learn about their anatomy and physiology. Often animals that have been mishaps of accidents or killed accidentally or even intentionally by unthinking people can be utilized as specimens for these studies. While the anatomy of the different species in a particular order or suborder does not differ greatly, really few reptiles have been studied in detail.

4. Identification specimens. Reptiles found dead in the field or captives that die even though given good care can be preserved as identification specimens in a collection for a school or a museum. Small lizards and snakes, for example, can be preserved by first immersing them in a 10 percent formalin solution. The body, limbs, and tail must either be injected with formalin or the skin must be slit to assure preservation. The formalin will in time discolor the specimens, but this can be counteracted to some degree by adding a pinch of baking soda to each cup of formalin solution. Or after several days in the formalin, specimens can be rinsed in water and then put in a solution of 75 percent ethyl alcohol or 35 percent methyl or isopropyl alcohol.

Each specimen should be tagged with a waterproof label that can be tied with a strong thread around an appendage or, in the case of limbless specimens, around the body. Data printed on the tag with a permanent ink should give the date and locality of capture and, if space, the collector's name or initials. Often a number is put on the tag referring to a catalog where these data and other information are recorded.

KEEPING LIVE REPTILES

Many kinds of reptiles make interesting pets, but keeping them is not encouraged unless they can be given the attention they need. It is far better to keep them only long enough to make particular kinds of observations and then release them where they were found.

Before keeping any reptiles, even temporarily, check with your local game and fish department or comparable agency to learn whether it is legal to do so. Regulations differ from state to state and also change frequently. If your study will contribute to basic knowledge, you may be issued a special permit and asked for a report on what you learn.

You may get your captive specimens as a result of finding eggs and then hatching them by keeping them in a warm, damp (not wet) place. You will not want more than one or two as captives. Release the others immediately in their natural habitat. Again, notify conservation officials who may want the young in a particular place.

Alligators

Baby alligators were once sold by the thousands to tourists for pets, and it is doubtful that more than a fraction of a percent ever survived a normal life span. Alligators do not make good pets no matter what their size or the circumstance. When it became illegal to sell baby alligators as pets, the smaller but similar (indistinguishable to tourists) Spectacled Caimans imported from South America were sold as baby alligators. Now it is illegal to sell any crocodilian as a pet, fortunate both for the animals and for people.

Lizards

Because they are more varied both in habits and in appearance than are snakes and also because they are mostly active during the day, lizards can be extremely interesting to watch in captivity. Catching a lizard is in most cases an exercise in agility. Sometimes you can attract the lizard's attention to one hand by waving or moving it while you move your other hand slowly and steadily forward to make the catch. But never grab a lizard by the tail (see p. 219).

Most lizards do well in captivity if you take care in providing for their needs. The smaller species can be kept in a terrarium (a converted glass aquarium is excellent) that duplicates as nearly as possible what the lizard would have in its natural environment—moist soil, a rotting log, and a rock for a woodland species, sand and xerophytic plants for desert species. Avoid spiny cacti, however, for in captivity lizards do not seem to be able to avoid sticking themselves. Make certain the lizards are provided with moisture. Some captive lizards do not drink water, and so if you spray the plants and other objects in the terrarium with water every few days, you will be giving them all they need. Moist foods are also a help. But even if you are convinced that the lizard does not drink, always provide a shallow dish of water.

Most lizards eat insects; a few kinds are vegetarian. Give your lizard the proper diet according to the species you have. Crickets, meal-worms, cockroaches, fruit flies—these are not difficult to get. Most lizards will also eat bits of dog or cat food.

Among the most commonly kept (and mistreated) are the horned lizards of the Southwest. They are sold, though it is illegal to do so now in most areas, as "horned toads." Horned lizards do not live long in captivity even though they may seem to be eating well and to have been provided with all of their needs. Because they are desert animals, they must be kept warm, even at night, and this may call for a separate heater to make certain the temperature never drops below 70 degrees F. (21 degrees C.). Raise the temperature to 80–85 degrees F. (26–29 C.) during the day.

Anoles that live in the southeastern United States are sometimes sold as "chameleons" because of their ability to change their color from brown or gray to green and vice versa. Courting males fan out their throat—a bright orange or red disc—and bob their body up and down. With good care, an anole may live for several years in captivity, proba-bly even longer than it would survive in the wild. But most people soon tire of their pets. The lizards then die. Because anoles are extremely active and are good climbers, they need a rather large cage or terrarium, which should be airy. They need live insects but may also eat cat or dog food. Exposure to a sun lamp for 20 minutes every day is desirable for diurnal reptiles.

Fence lizards, skinks, iguanids—all can be kept as captives. If you satisfy their dietary needs and give them comfortable surroundings, they will in turn perform much as they do in nature and will give you an opportunity to observe their behavior closely.

Snakes

As a result of myths, folklore, and lack of understanding, many people have a genuine fear of snakes. Seeing or being around a pet snake can help them in overcoming these fears, and this may ultimately be of benefit to snakes as a group. Pet snakes should not be used to scare people, of course, and no poisonous snakes should be kept as captives, even for short intervals, unless the cage is very secure and there is a special reason for having these dangerous animals.

Most harmless snakes will hiss, strike, and even bite when first captured, but if handled gently (hold them by the middle of the body, supporting all parts, never by the tail), they generally get over their nervousness rather quickly. In the field they can be carried in muslin bags or in mesh onion or potato bags (just be sure there are no holes). Never put these bags in the direct sun or in a closed car. Snakes succumb quickly to overheating.

If you do not expect to keep the snakes long, a cardboard box with a tight lid will serve as a temporary container. If you plan to have them for more than a few days, small snakes can be kept in a glass aquarium converted into a terrarium. Make sure it has a tight-fitting lid, for snakes are skilled escape artists. For more permanent housing, a wooden cage is best—that is, three sides and the floor of wood. One side can be glass so you can see in. The top can be screen. To give circulation, bore a number of small holes in the wood or cut one or two large openings and cover them with screen. Keep the cage out of direct sunlight—especially the glass side, as the glass becomes a magnifier that can turn the cage into an oven.

The less you have in the cage, the easier it will be to keep it clean. Snakes do like places to hide, however, and a piece of a log or a rock will generally suffice. For snakes that climb, provide a small limb with some branches. Keep the floor of the cage nearly bare. Snake excrement is semiliquid. It dries rapidly and can then be scraped up. Snakes drink regularly, and so always keep water in the cage. At shedding time, they may also soak in the water to help loosen their skin. This is particularly important for captive snakes because their diets are not exactly what they get in nature.

Most snakes like live foods, and unless a captive begins to eat soon, it should be turned loose. Force-feeding or liquids poured down the gullet through a funnel are possible but should be resorted to only if keeping the snake is important for some sort of study. Knowing the kind of snake you have is important in giving it acceptable meals. Water snakes like small fish, but it is apparently the smell of the fish that attracts them. Strips of beef or other meats ordinarily rejected are often accepted if rubbed with a fish to get the odor. Most snakes that kill by constriction, such as kingsnakes, will eat only warmblooded animals. This calls for live birds or mammals. But be careful. Even a large snake not in the mood to feed can actually be killed by a mouse!

Captive snakes commonly get mites under their scales. These can generally be killed by dusting the snake with a dog or cat flea powder. Put the dusted snake in a different container for an hour or two while you clean and disinfect its cage thoroughly. Often it is effective to put a small piece of a "pest strip" on the wire top of the cage, but never allow a reptile to come in contact with the strip. Mouth infections are also fairly common in captive snakes. Sulfa compounds or an aqueous solution of zephiran chloride can be used as a treatment, but if the condition persists get the advice of a veterinarian. Try to get the snake restored to good health before you turn it loose where it must fend for itself.

Every species actually presents a different kind of challenge and requires different accommodations. Tiny Ringneck Snakes, for example, can be made comfortable in a small terrarium. Black Rat Snakes, in contrast, need a large cage. Rat Snakes are mostly docile, while Racers and several other species are highly nervous and never seem to adjust to being captives.

No snake or any other reptile should be kept long. Remember that they become dormant in winter unless kept inside and that they are also unable to tolerate excessive heat. Duplicating their needs is difficult. This is something that should be left to zoos or similar institutions equipped to accommodate animals. Keep the snake as comfortable as possible, and then return it to nature where it belongs.

Turtles

Turtles that live in water are the most wary and difficult to catch, but if you persist, it can be done. Sometimes aquatic turtles are found wandering far enough from water to make catching them easy. Both the snappers and the softshell turtles have nasty dispositions and must be handled with great care to keep from being bitten. They can be kept in captivity only if you have a very large aquarium that can be modified so the turtles also have a place to rest out of water. Most of the semiaquatic turtles are easily kept in captivity, too, but many that are sold as pets are literally drowned because no place is provided for them to get out of the water for resting and breathing.

The most easily kept turtles are the land dwellers—the box turtles and gopher tortoises. All they really need is a fenced-in area in a yard, though they will also tolerate a terrarium or a cage. Because they do like to roam, they should be allowed to wander outside from time to time if they are ordinarily kept in confinement. Gophers can dig and will even burrow under a fence if not watched closely. This can be prevented if the enclosure is strictly theirs by burying the fence about six inches (150 mm) and also turning it under—that is, bent inward toward the enclosure. Turtles are not good climbers, and so the fence does not have to be high.

If turtles are kept inside, they will remain active all winter. If they are outside, they will become dormant in winter and must be provided with loose debris in which they can bury themselves. Aquatic or semi-aquatic turtles will bury themselves in debris at the bottom of a pool, which can then be filled in with leaves to provide a warm winter blanket needed for survival. A turtle from a southern area must never be left outside for the winter in a northern area.

Commercial turtle foods are fine for filling a turtle's stomach, but most do not provide complete diet needs. Bits of dog or cat food, or small fish, plus occasional earthworms and flies or other insects are much more adequate. As a generality, young turtles are almost exclusively carnivorous. Older turtles are at least partly vegetarian (exceptions are snappers and softshells, which remain carnivorous) and will take bits of fruits, lettuce, and similar foods. All of the aquatic turtles eat their food only underwater. Land turtles will eat from a dish. Most turtles will learn to get their food at the same place each day, but some will take food only when it is put directly in front of their nose. Some will eat food from your hand.

Eye infections and fungus growths are common illnesses of captive turtles. Respiratory illnesses occur in aquatic turtles kept in water that is too cool. These are generalities, of course. If it is necessary for you to keep a turtle (or turtles), consult a veterinarian to get the best treatment for any illness. After your observations are completed, give the turtles their freedom in their natural habitat.

Until laws were passed to stop the practice, countless thousands of baby sliders from southern United States were marketed as pets. The worst offense was painting their shells. This prevented a natural growth of the shell, causing first a deformity and eventually resulting in the turtle's death.

Again, keeping reptiles in captivity is possible, but the duration of their captivity should be brief. The purpose for keeping them should be to learn more about their lives and habits to make the living better for those still free in the natural world.

OTHER SOURCES OF INFORMATION

A nearby museum or college is one of the best sources for more information about reptiles. In addition, there are a number of organizations that will answer inquiries:

Division of Herpetology
Museum of Natural History
University of Kansas
Lawrence, Kan. 66045

Division of Herpetology
Field Museum of Natural History
Roosevelt and Lake Shore Drive
Chicago, Ill. 60605

Division of Herpetology
Museum of Vertebrate Zoology
University of California
Berkeley, Cal. 94720

Division of Herpetology
American Museum of Natural History
Central Park West at 79th Street
New York, N.Y. 10024

Division of Herpetology
National Museum of Natural History
Washington, D.C. 20560

There are three national professional organizations: the American Society of Ichthyologists and Herpetologists, the Herpetologists' League, and the Society for the Study of Amphibians and Reptiles. Each publishes one or more journals. If you are interested, inquire about these organizations at one of the addresses above.

The following publications also offer more information:

Ashton, Ray E., Jr., Stephen R. Edwards and George R. Pisani. *Endangered and Threatened Amphibians and Reptiles in the United States.* Society for the Study of Amphibians and Reptiles, Miscellaneous Publications, Herpetological Circular No. 5. Lawrence, Kan., 1976.

Collins, Joseph T., et al. *Standard Common and Current Scientific Names for North American Amphibians and Reptiles.* Society for the Study of Amphibians and Reptiles, Miscellaneous Publications, Herpetological Circular No. 7, Lawrence, Kan., 1978.

Conant, Roger. *A Field Guide to Reptiles and Amphibians of Eastern and Central North America.* 2nd ed. Boston: Houghton Mifflin, 1975.

Czajka, Adrian F., and Max A. Nickerson. *State Regulations for Collecting Reptiles and Amphibians.* Milwaukee Public Museum, Special Publications in Biology and Geology, No. 1, 1974.

Dowling, Herndon G., and W. E. Duellman. *Systematic Herpetology.* New York: American Museum of Natural History, 1979.

Ernst, Carl H., and Roger W. Barbour. *Turtles of the United States.* Lexington: University Press of Kentucky, 1972.

Fitch, Henry. *Reproductive Cycles in Lizards and Snakes.* University of Kansas Museum of Natural History, Miscellaneous Publications, No. 52, 1970.

228

Frye, Fredric. *Husbandry, Medicine and Surgery of Captive Reptiles*. Bonner Springs, Kan.: Medicine Publishing Co., 1973.

Gans, Carl, et al. *Biology of the Reptilia*. New York: Academic Press, 1970 to date (10 vols., continued).

Goin, Coleman J., Olive B. Goin and George R. Zug. *Introduction to Herpetology*. 3rd ed. San Francisco: W. H. Freeman, 1978.

Greenberg, Neil, and Paul D. MacLean. *Behavior and Neurobiology of Lizards: An Interdisciplinary Colloquium*. National Institute of Mental Health, DHEW Publication ADM77-491. Rockville, Md., 1979.

Grzimek, Bernard. *Animal Life Encyclopedia*. Vol. 6: "Reptiles." New York: Van Nostrand, 1975.

Guggisberg, C. A. W. *Crocodiles: Their Natural History, Folklore and Conservation*. Harrisburg, Pa.: Stackpole Books, 1972.

Harless, Marion, and Henry Morlock. *Turtles: Perspectives and Research*. New York: John Wiley and Sons, 1979.

Kauffeld, Carl. *Snakes: The Keeper and the Kept*. Garden City, N.Y.: Doubleday, 1969.

Leviton, Alan. *Reptiles and Amphibians of North America*. Garden City, N. Y.: Doubleday, 1970.

Milstead, William W. *Lizard Ecology: A Symposium*. Columbia: University of Missouri Press, 1967.

Neill, Wilfred T. *The Last of the Ruling Reptiles: Alligators, Crocodiles, and Their Kin*. New York: Columbia University Press, 1971.

Neill, Wilfred T. *Reptiles and Amphibians in the Service of Man*. Indianapolis: Bobbs-Merrill, 1974.

Oliver, James A. *The Natural History of North American Amphibians and Reptiles*. New York: Van Nostrand, 1955.

Parker, H. W. *Snakes of the World: Their Ways and Means of Living*. New York: Dover Publications, 1977.

Porter, Kenneth R. *Herpetology*. Philadelphia: W. B. Saunders, 1972.

Schmidt, Karl P., and Robert F. Inger. *Living Reptiles of the World*. Garden City, N.Y.: Doubleday, 1975.

Shaw, Charles E., and Sheldon Campbell. *Snakes of the American West*. New York: Alfred A. Knopf, 1974.

Smith, Hobart M. *Handbook of Lizards: Lizards of the United States and Canada*. Ithaca, N.Y.: Comstock Publishing, 1946.

Stebbins, Robert C. *A Field Guide to Western Reptiles and Amphibians*. Boston: Houghton Mifflin, 1966.

Wright, Albert Hazen, and Anna Allen Wright. *Handbook of Snakes of the United States and Canada*. 2 vols. Ithaca, N.Y.: Comstock Publishing, 1957.

INDEX

Individual species names, both common and scientific, are indicated with the text page only, since the text, illustrations, and range maps pertaining to each species run side by side. Although this is primarily a species index, there are also entries for each order, suborder, family, subfamily, and genus.

To find common names for species, look under the general headings "Turtles," "Lizards," "Amphisbaenids," "Snakes," or "Crocodilians," as appropriate.

To find scientific names, look for the genus first, and then the species or subspecies.

For an explanation of the names used in this book, see page 12.

232

233

239

A B C D E F